Casey A. Wood

Lessons in the Diagnosis and Treatment of Eye Diseases

Casey A. Wood

Lessons in the Diagnosis and Treatment of Eye Diseases

ISBN/EAN: 9783337140564

Printed in Europe, USA, Canada, Australia, Japan

Cover: Foto ©berggeist007 / pixelio.de

More available books at **www.hansebooks.com**

A Menstruum.

HORSFORD'S ACID PHOSPHATE

This preparation has been found especially serviceable as a menstruum for the administration of such alkaloids as morphine, quinine, and other organic bases which are usually exhibited in acid combination.

The admixture with pepsin has been introduced with advantage when indicated.

The Acid Phosphate does not disarrange the stomach, but, on the contrary, promotes in a marked degree the process of digestion.

Dr. R. S. MILES, Glencoe, Minn., says: "I use it in a great many cases as a menstruum for quinine, when an acid is necessary."

Send for descriptive circular. Physicians who wish to test it will be furnished a bottle on application, without expense, except express charges.

Prepared under the direction of Prof. E. N. HORSFORD, by the

RUMFORD CHEMICAL WORKS, Providence, R. I.

Beware of Substitutes and Imitations.

CAUTION:—Be sure the word "Horsford's" is Printed on the label. All others are spurious. Never sold in bulk.

Lessons in the Diagnosis

AND

TREATMENT OF EYE DISEASES.

BY

CASEY A. WOOD, C. M., M. D.,

Formerly Clinical Assistant, Royal London Ophthalmic Hospital (Moorfields); Microscopist and Pathologist to the Illinois Eye and Ear Infirmary; Professor of Ophthalmology, Post Graduate Medical School; Oculist and Aurist to the Alexian Bros. Hospital, Chicago.

WITH NUMEROUS WOOD-CUTS.

1891.
GEORGE S. DAVIS,
DETROIT, MICH.

To my friends,

WM. LANG, F. R. C. S.,
Surgeon Royal London Ophthalmic Hospital,

AND

DR. W. UHTHOFF,
Professor of Ophthalmology in the
University of Marburg.

TABLE OF CONTENTS.

Preface XI

LESSON I.

THE NORMAL EYE.

PAGE.

Introductory—A Plea for the General Study of Ocular Diseases—Some Simple Methods of Examining the Eye—Inspection of the Normal Eye as a Preliminary to Pathological Studies—Appearance of Healthy Eyelids, Conjunctiva, Sclera and Cornea—The Pupil, the Iris, and the Lens—Normal Tension—Focal Illumination—Exploring the Orbit—The Superficial Bloodvessels of the Eye—Testing the Visual Acuity and Muscular Power—Color Perception—Hints about Examining the Eyes of Children................... 1

LESSON II.

THE EYE IN DISEASE.

Examination of the Patient—The History of the Case—Signs and Symptoms—The Visual Acuity—Everting the Lids—The Bifocal Illumination—The Examination by Reflected Light—Some Points in Diagnosis—Abnormal Tension and Lacrymation—The Ocular Blood Supply in Disease—Remedies Commonly Used in Ophthalmic Practice—Leeches, Natural and Artificial—Heat and Cold—Blisters—Atropine, Homatropine, and Duboisine—Eserine and Pilocarpine—Cocaine—Boracic Acid and Nitrate of Silver—Other Remedies—Bandages and Shades—Goggles and Colored Glasses—Eye Cups and Eye Droppers—Irrigation of the Eye—The Cautery—Preparing for Operations—Antiseptic Precautions—A Clean Surgeon, Clean Instruments, and a Clean Patient...... 16

LESSON III.

DISEASES OF THE EYELIDS AND CONJUNCTIVA.

Atropine Irritation — Pterygium — Pinguecula — Lithiasis—Blepharitis—A "Black Eye"—Herpes of the Lid—Stye, or Hordeolum—Chalazion, or Cyst of the Eyelid—Lupus and Epithelioma—Xanthelasma—Entropion and Ectropium—Surgical and Other Treatment of These Conditions—Hotz's Operation—Ptosis—Lagophthalmos—Symblepharon.............. 32

LESSON IV.

DISEASES OF THE LACHRYMAL APPARATUS.

Epiphora—Lacrymation—Dacryocystitis-Lachrymal Abscess—Slitting Up the Canaliculus—Passing the Nasal Probe—Treatment of Lachrymal Obstruction by the Syringe of Meyer or Anel..................... 49

LESSON V.

DISEASES OF THE CORNEA AND SCLEROTIC.

Arcus Senilis—Keratitis—Phlyctenular Keratitis—Foreign Bodies in the Cornea—Interstitial Keratitis—Punctate keratitis—Ulcers of the Cornea—Spreading and Non-spreading Ulcers—Hypopyon and Onyx—Paracentesis of the Anterior Chamber—Leucoma Adherens—The Use of the Cautery—Senile Ulcer—Opacities of the Cornea—Nebula, Macula, and Leucoma—Staphyloma Anterius—Tatooing the Cornea—Scleritis and Episcleritis............................. 56

LESSON VI.

DISEASES OF THE IRIS AND ANOMALIES OF THE PUPIL.

Coloboma of the Iris—Differences in Color—Albinism—Nystagmus—Iridodialysis—Various Kinds of Iritis—

ix

PAGE.

Iridectomy — Variations in the Size of the Pupil — Myosis and Mydriasis — Various Conditions which Produce Anomalies of the Pupil — Hippus.......... 79

LESSON VII.

CATARACT AND OTHER AFFECTIONS OF THE CRYSTALLINE LENS.

Dislocation of the Lens — Iridodonesis — Aphakia, or Absence of the Lens — Cataract — Nuclear and Cortical Cataract — Senile Cataract — The Operation for Removal — When to Operate — After-treatment — Complications — Soft Cataract — The Zonular or Lamellar Cataract of Children — Discission or Needling....... 97

LESSON VIII.

GLAUCOMA.

The Necessity of an Early Diagnosis — Varieties — Causes — Intraocular Changes in the Disease — Signs and Symptoms — Treatment — Iridectomy and Sclerotomy — Eserine................................ 112

LESSON IX.

OCULAR AFFECTIONS IN GENERAL DISEASES.

Manifestations of Syphilis, Rheumatism, and Other Diatheses — Muscæ Volitantes — Amblyopia — Toxic Amblyopia — Eye Symptoms in Tobacco and Alcohol Poisoning — Abscess of the Orbit — Graves' Disease — Progressive Locomotor Ataxia — Diphtheria — Bright's Disease — Migraine — Malaria — Reflex Neuroses — Sympathetic Ophthalmia — Penetrating Wounds of the Globe — Sympathetic Irritation and Inflammation — Treatment — Enucleation of the Eyeball................................... 117

LESSON X.

PARALYSIS, SQUINT, AND OTHER MUSCULAR TROUBLES.

The Physiology of the Subject—The Nerve Supply—Ocular Paralyses—Their Symptoms—Paralysis of the Sixth Nerve—Paralysis of the Fourth Nerve—Oculo-motor Paralysis—Ophthalmoplegia—Causes and Treatment of Paralysis—Strabismus or Squint—Convergent and Divergent Squint—The Measurement of Squint—Treatment—Operations for Strabismus—Tenotomy—Advancement 134

INDEX 147

PREFACE.

The purpose of this manual is to aid the physician to detect and treat, by means always at hand, those diseases of the eye which experience has shown are most frequently overlooked in the course of general practice. Ocular diseases which are commonly and easily diagnosed by the non-specialist are not so much dwelt upon as those that are more obscure; while two subjects of great general importance, already fully and ably discussed in this Series,—Conjunctival Diseases, by Prof. Mittendorf, and the Determination of the Necessity for Wearing Glasses, by Prof. St. John Roosa—although partly falling within the scope of such a treatise as this, will be passed over. A perusal of these monographs would be necessary to complete the programme just laid down.

Frequent references, for the student's benefit, will be made not only to the pages of this but to those of the other works on the eye published in the Series. To save space, these references will be made by bracketed page and initial only. For example, (27) means page 27 of this book; (R. 34) will refer to page 34 of Prof. Roosa's work on the Determination of the Necessity for Wearing Glasses; (M. 40) is intended to indicate page 40 of Prof. Mittendorf's monograph on Granular Lids and Contagious Ophthalmia; while (C. 22) means page 22 of Dr. Claiborne's book on the Theory and Practice of the Ophthalmoscope. Of course, this manual makes no pretension to being a *complete* treatise upon the subjects of its chapter headings. The writer trusts, however, that the space at his disposal has been filled with such practical hints and useful information as will be most likely to assist the non-specialist practitioner.

204 Dearborn St., Chicago.

LESSON I.
THE NORMAL EYE.

Introductory—A Plea for the General Study of Ocular Diseases—Some Simple Methods òf Examining the Eye—Inspection of the Normal Eye as a Preliminary to Pathological Studies—Appearance of Healthy Eyelids, Conjunctiva, Sclera, and Cornea—The Pupil, the Iris, and the Lens—Normal Tension—Focal Illumination—Exploring the Orbit—The Superficial Blood-vessels of the Eye—Testing the Visual Acuity and Muscular Power—Color Perception—Hints about Examining the the Eyes of Children.

The conduct of the busy practitioner towards such ocular affections as occur in his practice usually differs from his treatment of most other diseases. The acquirement of even a moderate degree of knowledge respecting diseases of the eye—especially those requiring the use of the ophthalmoscope—necessitates considerable study and frequent chances for observation. The every-day routine of practice affords neither opportunities for the one nor time for the other, and it is not, therefore, surprising that such cases are usually referred to a *confrère* who makes a special study of them.

While one must admit the advantages that, on the whole, accrue to both patient and medical man from such a disposition of this class of maladies, it is not without its drawbacks.

There is a fairly well-defined and important

group of diseases, affecting various parts of the visual apparatus, whose signs and symptoms are especially prone to be overlooked or misinterpreted by the general practitioner simply because he has abandoned the inspection of the eye. In this group are some that call for recognition in tones so loud that they ought to be detected at once by any man who will use his ordinary powers of observation, while others require stricter attention upon the part of the observer. Many of them, too, are by no means simple matters. When not recognized and promptly treated by the surgeon, they go on, in the ordinary course of events, either to a more or less rapid destruction of the organ itself, or to considerable impairment of its function. The claims which the study of this particular set of diseases makes upon the profession become all the more urgent when one reflects that in every instance an early diagnosis may be correctly made by the exercise of the same quality and amount of care and intelligence which are commonly brought to the investigation, let us say, of affections of the lungs and uterus.

We shall accordingly proceed to consider this group of maladies entirely from the standpoint of the physician in general practice, who, it is assumed, will be confused rather than helped by the introduction of references to ophthalmoscopic findings, perimetric measurements, and other subjects of interest usually comprehensible by the specialist alone.

EXAMINATION OF THE NORMAL EYE.

Just as one studies the conditions under which the healthy lungs and heart subserve their respective functions before proceeding to an investigation of the departures from health to which these organs are subject, so it is well to make a study of the normal eye. The physician should cultivate this habit of observation in the case of the visual apparatus particularly, because so much information as to its condition—whether normal or abnormal—can always be learned by SIMPLE INSPECTION.

To be of the greatest value, such an examination should be methodical, and every case should be examined in the same order. Of course, where the patient's condition is so obvious that it would be a mere waste of time to follow out a programme including all the parts of the eye, this rule may be modified; but it is safest to run over the whole ground, so far as possible, in the great majority of instances.

The common, and probably the best, plan that can be adpoted is that of proceeding from without inward, bearing in mind, meanwhile, as many as possible of those diseased conditions which one might expect to find in the particular locality under examination.

Look, then, for a moment at a pair of normal eyes.* When they are directed straight forwards the

---
*Only such references will be made to the anatomy and physiology of the eye and its accessories as are familiar

EYELIDS are equally separated and expose nearly the whole of the circle of the iris. Their edges are smooth, regular and of uniform color throughout and are fringed by a single row, or by evenly distributed rows, of lashes (*cilia*) all of which point away from the underlying parts which they are designed to protect. Each cilium can be traced directly to its little opening in the edge of the lid, and the space between it and its neighbor harbors no swelling, dried secretion or other abnormal product.

The PALPEBRAL SKIN has the appearance of healthy skin elsewhere upon the face and it is loose enough to admit of free movement of the lids, which are capable both of being tightly closed and of being opened to their fullest extent.

At the INNER CANTHUS is the CARUNCLE, a reddish white prominence resulting from a reduplication of the conjunctiva. Encircling it, but running in the substance of the lid margins, are the two CANALICULI —upper and lower—whose external openings (the *puncta*) can readily be seen when the lids are slightly everted.

Note that the puncta are closely applied to the

---

to the general reader. More extended anatomical and physiological descriptions, as well as fuller statements in the pathology and microscopy of the subject, can be had in most text-books. The advanced student will also find Dr. Claiborne's book on the Theory and Practice of the Ophthalmoscope very helpful.

eyeball so that, by a combined drainage and suction action the tears are carried along the canaliculi into the LACHRYMAL SAC and then to the nose by the NASAL DUCT. In the normal state little or no fluid of any kind can be pressed out of the sac through the puncta.

The *lower lid is everted* by directing the patient to look upward while the skin near the palpebral margin is, with the forefinger, pressed downward and backward over the edge of the orbit. The *palpebral conjunctiva* is smooth, of a faint pink color, and is thrown into folds just where it extends forward to cover the eyeball. The tubules of the *Meibomian glands* can also be seen underneath the mucous lining. They discharge their fatty contents upon the free margin of the lid and assist in preventing an overflow of tears upon the cheek.

If the subject under examination be told to look downward and a probe, held in the left hand, be laid along the groove beneath the upper margin of the orbit, the *upper lid may be everted* and its mucous surface exposed and examined by grasping its cilia between the right thumb and forefinger and gently turning it over the instrument. This little manœuvre will be all the more successful if just before the moment of eversion the lid be pulled, in a direction forward and downward, away from the globe to which it is applied.

The meibomian tubules are now more easily and

distinctly seen than are those of the lower lid. The conjunctiva, like that of the lower lid, is smooth and transparent, so that we easily see the bloodvessels or other structures beneath it and in its substance. The outlines of the TARSAL CARTILAGE can be made out, while above the superior border of the latter are the "folds of transmission" corresponding to those already mentioned as occurring in the lower lid. These should be carefully examined by telling the patient to look down at his feet and at the same time putting the lid somewhat on the stretch. Not only are these folds of mucous membrane especially subject to disease, but they form a favorite hiding place for foreign bodies.

The OCULAR CONJUNCTIVA in its normal condition looks white because it is translucent and the glistening sclerotic shines through it. The bloodvessels which occupy a plane anterior to the latter, being opaque are consequently visible as long as the conjunctiva remains normal.

It must always be remembered that the bloodvessels of the conjunctiva proper run both in the submucosa and in the membrane itself, and that they move when the latter is moved. This fact is best demonstrated by "sliding" the ocular conjunctiva over the globe. Pick out a vessel on the healthy eyeball. Press upon it by means of a forefinger placed at the margin of the lid, the latter being between finger and vessel. If it moves *with* the displaced conjunctiva, it is a conjunctival vessel. If not, it must lie deeper than the latter.

The underlying SCLERA (sclerotic coat) presents an even, uniformly white and smoothly polished surface. Fitted into it at the sclero-corneal junction is the "watch glass" of the CORNEA. No vessels ramify on its surface or in its substance; it is as transparent as glass and forms an almost perfect circle. Place the patient opposite a window and tell him to follow with his eyes the uplifted finger (held about 18 inches from his face) which is moved in various directions.

The picture made by the window frame upon the cornea should be distinct and not broken or distorted, showing that the image is reflected from a perfectly smooth and regularly rounded surface.

The AQUEOUS HUMOR, filling the anterior chamber, is equally well adapted to the transmission of light, being as clear as water.

The ANTERIOR CHAMBER itself is regular and well defined—not too deep, not too shallow.

The IRIS has a peculiar hue and brilliancy. Note its soft, velvety appearance, and how closely it resembles its fellow of the opposite side. Its pupillary margin has the same velvety though well defined appearance, and nowhere is it attached to the cornea in front or the lens behind.

A superficial observer might imagine that the peripheral attachment of the iris is coincident with, or nearly coincident with, the sclero-corneal junction. It is well to notice, however, that outside of this junction there is quite a large angular space formed by the iris and sclera. In the substance of the latter and close to this angle lies the circular CANAL OF SCHLEMM.

Both PUPILS, round and nearly centrally placed, respond equally and readily to the stimulus of light. If a hand be placed over each eye, so as to entirely exclude the light, the pupils at once dilate, and if, after a minute's waiting, first one hand and then the other be removed, the irides contract quickly, oscillating somewhat until they reach a state of comparative rest. They should, under similar conditions, be of the same size.

Another important experiment is the following: The eyes are directed (in a subdued light) for a minute to a distant object and then suddenly fixed upon another object a few inches in front of the nose. A decided contraction of the pupil takes place—a contraction which is followed by dilatation, on again looking in the distance. This is the action of the pupil to ACCOMODATION (R. 14).

Telling the individual under examination to look down at his feet (so as to present the superior sclerotic surface), place the tips of both forefingers on the upper lid at the centre of the upper orbital margin. Gently press the globe, first with the tip of one finger and then with the other. It yields a firm, yet elastic impression. This is the NORMAL TENSION of the eye and by a little practice the physician may soon be able to decide whether, in a given instance, the tension is more resisting or is softer than normal. In the first case it would be indicated by a plus sign (+1, +2, etc.), according to the degree of increase; in the second instance by a minus sign (—1, —2, etc.) to indicate the amount of lessened tension. *Also, note that moderate pressure upon this and other parts of the eyeball is painless.*

In youth the normal crystalline lens is invisible, but in old people it becomes slightly hazy, and may usually be seen by concentrating the light from a lamp or gas jet (placed at the side of the patient's face) upon it by a convex lens having a focus of from 5–8 centimetres (2–3 inches). Such a glass is readily obtainable from any optician or instrument maker, and is indispensable in making a proper examination of the eye. By its aid this method of OBLIQUE ILLUMINATION, as it is called, can be applied to searching for foreign bodies and opacities in the cornea, changes in the iris, opacities in the lens and lens capsules, and to many other uses—as we shall hereafter see.

The glass is held between the thumb and index finger, the hand being steadied by touching with one or more fingers the forehead or cheek of the patient, and the rays of light are directed sideways (*die seitliche Beleuchtung* of the Germans) and brought to a focus on the part of the eye to be examined. This point of light may be made to play all over the external eye and even to illuminate growth, foreign bodies, clots, etc., situated as far back as the vitreous body, behind the lens.

Under normal conditions neither globe should protrude between the lids, nor should one eye be more prominent than the other, although it must not be forgotten that many persons exhibit a shallow facial development which gives to their perfectly normal eyes the appearance of having been pushed forward from behind.

In this connection it is well to become acquainted with the impression given to one's fingers—little fingers to be preferred—in EXPLORING THE ORBIT. Note the position of the pulley of the superior oblique muscle, the canthal ligament, and the notches or foramina for the infra- and supra-orbital nerves, so that they may not be mistaken for abnormalities.

Some attention should be given to the BLOOD SUPPLY of the superficial portions of the visual apparatus. The movable vessels of the conjunctiva have already been noticed. It is well to note their number, their size, and the fact that they do not affect the coloration of the mucous membrane to which they are supplied. Then there are in addition to these a few rather large and tortuous vessels (*anterior ciliary arteries*) which lie in the ocular sub-conjunctival tissue and run forward on the globe until, at a point one or two millimetres from the cornea, they pierce the sclerotic, and supply it, the iris, and the ciliary body. These arteries give off small and nearly straight (*episcleral*) branches, which form a close network of vessels around the corneal margin. *They are invisible in health.* When visible (and not moving when the ocular conjunctiva is slid over the sclera), they point to irritation or inflammation of the iris, cornea, or some deeper structure.

The EQUILIBRIUM OF THE OCULAR MUSCLES (R. 47), may, in normal eyes, be tested as follows: Having covered with the left hand one of the person's eyes, tell him to "fix," *i. e.*, look steadily at, the tip of a pencil or other small object held in the right hand about ten inches in front of his nose. Then quickly transfer the left hand to, and cover up, the other eye, watching closely the recently uncovered eye. When the various muscles of the eye are properly balanced, the uncovered eye will not turn out or in, indeed it will not be moved at all, to fix the object held in front of the face. On the other hand, there is muscular weakness present (which not infrequently gives rise to troublesome symptoms) should the second eye move inwards or outwards in its efforts to see distinctly the object in front of it.

Also, both eyes should, first *singly* and then *together*, follow the movement of an object moved to and fro, upward and downward, in front of the eye, until portions of the cornea are covered. That is the *excursion* of the normal eyeball.

Having in this way made an *objective* examination of the normal eye, it is advisable to investigate some SUBJECTIVE PHENOMENA.

First, the VISUAL ACUITY (R. 2). This is nothing more than the ability which the eye possesses of recognizing the *form* of objects. To make a distinct impression upon the retina, objects must be of a certain size which is found to bear a definite relation to their distance from the eye—the nearer they are the smaller they *may* be; the farther away the larger they *must* be. The plan commonly adopted is to choose letters of the smallest size which can be readily distinguished by the normal eye at 6 metres (20 feet). These letters in the ordinary series (Snellen's test types) are marked $\frac{6}{6}$ or $\frac{20}{20}$, indicating that the eye which reads them at this distance—6 metres, or 20 feet—has the full acuteness of vision. If from the same distance the letters (twice as large as those marked $\frac{6}{6}$) marked $\frac{6}{12}$ or $\frac{20}{40}$ are the smallest that can be read, the person's vision is only one-half the normal. And so on through the whole series down to $\frac{6}{80}$ or $\frac{20}{200}$, the numerator of the fraction always designating the distance from the test types at which the patient stands, the denominator the distance at which the test line should be distinguished by the *normal* eye. It may, however, happen that the patient cannot read even the largest letter of the test types. In that case he should gradually approach until he is able to distinguish it, and the number of metres (or feet) distant from the types, divided by 60 (or 200), will give the amount of his visual power. For example: A patient distinguishes with his right eye the largest letter of Snellen's types at 2 metres (6½ feet). His left eye, however, can read at 6 metres the line marked $\frac{6}{30}$ or $\frac{20}{100}$. Shortly expressed, we would say: Right Eye (or R.) Vision (or V.) = $\frac{2}{60}$ (or $\frac{6}{200}$); L. V. = $\frac{6}{30}$ (or $\frac{20}{100}$). But the ability to distinguish form may be further reduced, and then we find out at what distance from his face the patient can count one's fingers. Here V. = fingers at 10 inches, or whatever the distance may be. Finally, the patient may be so blind as to able to distinguish between light and darkness only, when V. = p. l., or perception of light. In all these cases one eye at a time is examined, the other being covered with the hand or a light bandage which does not press upon the eye. Snellen's test types are readily obtainable from any medical bookseller, and are very useful in the diagnosis of diseases of the eye.

In addition to the use of these test types—which are employed for examining distant vision only—there are *tests for near vision* (R. 15.17). Those commonly employed are Snellen's and Jaeger's. The latter correspond to ordinary type, and are numbered from 1, the smallest, to 20, the largest. Persons who have full *distant* visual acuity are usually able to read the finest print, and, *vice versa*, those who are able to read Jaeger No. 1 have good distant vision. The exceptions to both these statements are numerous and cannot be considered here.

PERCEPTION OF COLOR is one of the functions of the healthy eye, a function which (congenitally deficient in 3 per cent. of all persons) becomes impaired or abolished in certain diseases. The normal eye should immediately recognize the various colors and shades of color. Samples of Berlin wools make the best test objects.

The EXAMINATION OF THE EYES OF CHILDREN requires the exercise of considerable tact combined with the greatest care and gentleness.

Except in special cases, to be spoken of later, most information can be obtained by drawing the little one's attention to some toy which is moved in various directions in front of its face. This will, in most instances, enable the observer to see the cornea, the iris, and the lens, and permit a fairly satisfactory examination of the external eye to be made. When there is much photophobia, or the lids are swollen, or where from any cause the child is unable or unwilling to open its eyes, it may be necessary, before a proper examination can be obtained, to drop into the eyes a few drops of a weak (2 per cent.) solution of cocaine, and then to open them by means of lid retractors, a pair of which should form a part of every practitioner's *armamentarium*. It is in every way most satisfactory to place the little one's head between the operator's knees, over which a towel has been thrown. The nurse holds the child lying on her lap, and grasps the hands in one of hers, while the surgeon has both his hands free to use the retractors, apply remedies, etc. It is sometimes necessary to administer chloroform.

It is possible, although it requires some practice, to evert both lids in infants by pressure at the orbital margins. The palpebral skin is directed over the orbital margin backwards and inwards by the thumb or finger-nail, and the conjunctiva is effectually exposed. The crying which the child indulges in helps rather than hinders this manœuvre.

LESSON II.

THE EYE IN DISEASE.

Examination of the Patient—The History of the Case—Signs and Symptoms—The Visual Acuity—Everting the Lids—The Bifocal Illumination—The Examination by Reflected Light—Some Points in Diagnosis—Abnormal Tension and Lacrymation—The Ocular Blood Supply in Disease—*Remedies Commonly Used in Ophthalmic Practice*—Leeches, Natural and Artificial—Heat and Cold—Blisters—Atropine, Homatropine, and Duboisine—Eserine and Pilocarpine—Boracic Acid and Nitrate of Silver—Other Remedies—Bandages and Shades—Goggles and Colored Glasses—Eye Cups and Eye Droppers—Irrigation of the Eye—The Cautery—*Preparing for Operations*—Antiseptic Precautions—A Clean Surgeon, Clean Instruments, and a Clean Patient.

One should preserve the same order in examining a patient suffering from an ocular affection that is commonly followed in diseases of other parts of the organism. Having first noted the patient's age (66, 60, 99) obtain as complete a history of the case as possible. Has he, for instance, any other affection, local or general, which is likely to affect the diseased eye or which may be the chief cause of the trouble (117)? When did the ocular affection first appear? Did it come on slowly or suddenly (112)? Has it affected one eye or both (120)? If both are diseased when did the second one become involved (68)? To what cause does the patient or his friends attribute the disease? What signs and symptoms have been most prominent during the illness? Has there been any pain (57, 81)? If so inquire into its position, its character and its frequency. Does it grow worse at any particular period of the twenty-four hours (84)? Is there any "discharge" from the eyes? If so, does it cause adhesion of the lids? Has the patient observed specks or spots (20) floating in front of his eyes, any halos about lamps or gas jets (114), or does he occasionally notice sparks, or balls of fire, or colored light. Is the disease decreasing, increasing, or is it at a standstill? What treatment, if any, has so far been given?

These questions should, I think, be asked in the great majority of instances and the substance of the answers recorded in the case book which every careful physician will keep. Next, the examiner will rely upon his own observations and note carefully anything abnormal about the patient which he may think has any bearing on the ocular disease. These may be signs of syphilis, congenital or acquired (66), of tuberculosis of emaciation or its opposite, of pulsation in the cervical veins an abnormal pulse, flushed face and so on.

The successful ophthalmologist is ever alive to these important general signs of disease.

The visual acuity may at this stage be noted (12) as it is commonly desirable to see what interference with vision has been produced by the disease, and later on what improvement (if any) has resulted from the treatment. In applying this test where the eye is inflamed it is always well for obvious reasons to have the patient's immediate surroundings in partial darkness and to make a note of the fact so that subsequent examinations may be made under similar circumstances.

Now it is probably time to specialize and to proceed to an examination of the ocular region.

The more acute the disease (and especially if it be an affection of the palpebræ, cornea, or iris), the more decidedly are the tissues of the temple, brow, face and lids likely to be swollen and hot. The superficial vessels of these parts will then be distinctly seen running through the puffy skin.

The hot tears, mingled it may be with abnormal secretion, bathe the lower lids and escape at the outer and inner canthi and thus add another irritant to the inflamed parts.

The nasal duct is filled to overflowing with the lachrymal fluids and the patient finds it necessary to use his handkerchief freely.

Not infrequently a pad of cotton wool, enclosed in a not over clean bandage, is worn over the eye. This absorbs the altered secretion and is thus made to act as a kind of morbific poultice which further increases the discomfort of the patient by producing excoriations of the palpebral skin.

Our first duty is to cleanse such an eye with a

warm and mild antiseptic solution—say of boracic acid—gently applied with some aseptic cotton wool. If marked photophobia be present instilling a few drops of a two per cent.—10 grains to the ounce—solution of cocaine will, in ten or fifteen minutes, allow of a more comfortable and more thorough examination of the affected parts.

It is good policy to evert the lids in all such cases (5) when it is not specially contra-indicated, and study the condition of their ocular surface (M. 5, and 17). The trouble may be entirely there.

The cornea, iris and sclera should now be carefully looked over, first with ordinary illumination and then, if the condition of the patient will allow it, by means of the focal illumination (9). A good plan, when one wishes to explore carefully the details of a lesion of the cornea, sclera, or even the conjunctiva, is to proceed as above and then to use in addition, as a magnifying glass, a second lens held in the other hand. By this means a most accurate, though enlarged, picture of the part can be had. In the same way opacities of all kinds in the cornea, alterations in the iris, and most changes in the lens structure can be made out.

But we should have other methods of diagnosis. For determining the presence or absence of opacities in the aqueous, lens, and vitreous body, when an examination is not prevented by a too opaque cornea, it is a very good plan to examine these media in a dark room with the light from a single gas jet or lamp.

Both being seated, the examiner faces the patient, at the side of whose head is placed the light, and the surgeon throws its rays upon the pupil under examination by means of a small concave mirror, pierced with a central opening, held in front of his eye. The hole in the mirror should not be more than 3 m. m. in diameter. It is always justifiable in making this particular examination to dilate the pupil. A couple of drops of a two-per-cent. cocaine solution will do this; it is a harmless mydriatic whose effects pass off in a few hours. If the observer's eye be placed about 30 cm. in front of the patient, he will see that an uninterrupted reddish reflex (from the retina) has taken the place of the previously black pupil. Opacities of any kind in any of the media will then appear as small black objects in this red field. If the patient be now directed to look slowly up to the ceiling and then at the mirror, the surgeon will be able to judge, with a little practice, whether the objects change their places relative to their surroundings. If they do, they are probably in the vitreous. If not, they are probably in the lens. These two methods should be employed in the order named and will seldom fail to detect the presence of abnormal products in the optic media—a very common cause of disturbed vision.

Reference has already been made to the normal blood supply (10). The blood vessels of the eye are almost always affected in disease of the organ, and the surgeon should be careful to note in each case, as a means of diagnosis, to what extent the circulation has been disturbed.

Next the tension of the globe should be tested (9) and the examiner may inquire at the same time whether the pressure causes any pain—whether the parts are tender (9).

When lachrymation is a constant symptom (without definite signs of acute disease) pressure should be made over the lachrymal sac to discover whether by this manœuver any abnormal secretion—pus or mucopus—can be squeezed out (50).

REMEDIES COMMONLY USED IN OPHTHALMIC PRACTICE.

It is generally agreed that local blood-letting is a valuable agent for the reduction of most deep seated and acute inflammations of the eye (89). The artificial leech answers very well for this purpose. I much prefer it to the living animal, which is often uncertain in its action. Of course, the amount of blood to be withdrawn and the frequency of the application will depend upon the severity of the disease and the effect of the remedy. Very little effect is produced unless half an ounce of blood is withdrawn. The favorite spot for the operation is the temple half an inch from the outer margin of the orbit.

HEAT is often applied after or in conjunction with the leeching. This may be in the form of steam, hot water, or hot compresses. A good plan for applying the first named is to take a tumbler, heat it and fill it half full of boiling water. The mouth of the tumbler is then closely applied to the eye and brow, which may be thus well steamed for five or ten minutes at a time as often as necessary. If hot compresses are used, the applications should be made for the same length of time and at intervals.

The practice of poulticing or of using hot applications to the eye for hours at a time appears to me to be harmful instead of helpful.

When COLD is employed (64) pieces of flannel four inches square may be kept on a block of ice, the the lower one being changed when needed for the one just removed from the eye In the absence of an attendant, the patient may often do this himself.

BLISTERS are valuable adjuncts to eye surgery, when judiciously employed. They may be applied above the brow, at its outer edge, or to the temple.

ATROPINE is one of the most valuable drugs we possess. One drop of a one per cent. solution produces wide dilatation of the pupil in half an hour. This action begins fifteen minutes after it is introduced. Three hours afterwards the accommodation (R. 14) is fully paralyzed, and the effects do not pass off in some patients for a week; in others, a shorter time is required. It is well to say here that solutions of atropine and the other alkaloids used in ophthalmic surgery, if they do not actually deteriorate in time, become cloudy from the formation of moulds.

It is consequently advisable to have them made up with a solution (1:5000) of mercuric chloride. In my hands a saturated solution of boracic acid, used by many, has not prevented the growth of these forms of life.

HOMATROPINE resembles atropine in its effects upon the pupil and accommodation, except that it may be said to begin its action earlier and to run its course sooner than the former. In twenty-four hours after the instillation of a few drops of the one per cent. solution, the effects upon the accommodation have mostly passed away. In consequence of these qualities, it is employed for dilating the pupil when the ophthalmoscope is to be used.

DUBOISIA is a quick and powerful though not so lasting a mydriatic as atropine, and is employed (in about the same dose) instead of the latter when it produces irritation or inflammation of the conjunctiva and swelling of the lids.

Of the drugs which contract the pupil, myotics, which, generally speaking, have an opposite effect to agents of the atropine class, the most important is ESERINE.

This alkaloid is derived from the Calabar bean, and is also known by the name *physostigmine*. It is usually given in weaker doses than atropine (say ½ to 2 grains to the ounce) on account of the frontal pain and twitching of the eyelids which it causes. For this reason, also, it is sometimes combined with cocaine (4 grains to the ounce). The pain, though severe, soon passes off, and when it is found necessary to instil the drug for some length of time, less and less irritation is produced until, finally, its use is not followed by the first effects. Full contraction of the pupil with spasm of accommodation is brought about in fifty minutes. Recovery takes place in three days.

PILOCARPINE, an alkaloid obtained from jaborandi, is a myotic, but weaker in its action than eserine. It is, in addition, a powerful sudorific and expectorant, and in doses of $\frac{1}{8}-\frac{1}{2}$ grain is much used hypodermically in diseases of the eye—especially in choroiditis, and in inflammations and detachment of the retina.

If used locally, like eserine, a one per cent. solution is the usual strength.

COCAINE (from the *erythroxylon coca*) is a late but extremely valuable contribution to ocular therapeutics. With its aid many operations hitherto performed with the aid of ether or chloroform may now be undertaken while the eye is under its anæsthetic influence alone. When first dropped into the eye a two per cent. solution causes a little smarting. This passes off in a minute or two and the conjunctiva and cornea become completely anæsthetized in from four to six minutes.

Five minutes afterwards the numbness begins to pass off and the normal state is reached in another quarter of an hour. The alkaloid causes contraction of the blood vessels, whitens the sclera, dilates the pupils and slightly weakens the accomodation power. The mydriasis may remain for twenty-four hours. Cocaine greatly increases the mydriatic effect of atropine. In the same way mixtures of homatropine and cocaine are used for paralyzing the accomodation in making examinations for the correction of refractive errors. A temporary but very effectual result may thus be obtained by using one drop of a one per cent. solution in castor oil of these last named alkaloids.*

For operative purposes solutions of cocaine should be made fresh with distilled or boiled water. Not more than three or four instillations (within five minutes) are needed for cataract extraction, and a two per cent. solution is quite strong enough. Small tumors (chalazia, etc.) may under its influence be removed without pain if a 4 per cent. solution be injected under the skin or about the growth. Cocaine is also valuable when it is found necessary to apply caustic irritants to the external eye. Here it is well to paint over the spot to be burned with a 10 per cent. solution of the alkaloid. If freely used it causes dryness and loosening of the corneal epithelium. This may usually be prevented by ordering the patient to keep the affected eye closed.

---

*See author's paper in the Journal of Ophthalmology, Otology and Laryngology, July, 1889.

Lamellæ or discs. A very elegant method of preserving and applying these alkaloids is in the form of minute discs of gelatine. Several reliable chemists make them, and they combine the advantages of a portable form, definite dose and complete preservation. One of these small discs will adhere to a damp match, camel hair pencil or probe, and may be laid in the conjunctival sac, or against the sclera, the lower lid being meantime drawn down and the patient told to look up.

Numerous metallic salts are employed by ophthalmic surgeons. These are usually directed against affections of the conjunctiva, cornea and sclera. Probably the most commonly used of these is BORACIC ACID. A saturated solution contains about 20 grains to the ounce of water. This and weaker solutions give little or no pain when applied to the eye and they make mild and effective antiseptic lotions. The salt itself when applied to the eye is practically non-irritant. An ointment is also in common use.

BORAX (the biborate of sodium) has a weaker antiseptic action than the former, but, in about the same dose, is used as a cleansing lotion to the eye.

OXIDES OF MERCURY play an important role in ocular therapeutics. A common ointment is *Pagenstecher's*:

        Yellow oxide of mercury........ 24 grains.
        Vaseline or cold cream.......... 1 oz.

This makes a strong mixture and it is always well

to prescribe a much weaker one for the patient's use. The red oxide is used in the same proportion and for much the same purposes.

The ointment of the acid nitrate of mercury, CITRINE OINTMENT, is preferred by some surgeons to the foregoing. Instead of the neat's-foot oil, it may be made with cod-liver oil.

SOLUTIONS OF PERCHLORIDE OF MERCURY, in various strengths, are useful as antiseptic lotions, and are widely employed for cleansing the conjunctiva and eyelids previous to and after operations. For this purpose a strength of 1:5000 is about right.

CALOMEL in fine powder, blown into the eye with a blower, or flicked with a camel's-hair brush, is useful in chronic corneal diseases.

When IODOL or IODOFORM are employed in eye surgery, they are used as a strong ointment with vaseline or in the form of impalpable powder.

ZINC CHLORIDE or SULPHATE (½ to 2 grains to the fluidounce of water) is a valuable astringent.

ALUM, in stick or crystal, is a useful application in chronic forms of conjunctivitis. A lotion (4 to 10 grains to the ounce) is also used in mild forms of acute and chronic catarrh of the conjunctiva.

SULPHATE OF COPPER, in the form of pencils or of a smoothly pointed crystal, is a classical remedy in trachoma. It is a mild escharotic, and its use causes considerable smarting. A few drops of cocaine solution, 2 per cent., will relieve that. A good preparation to be used instead of bluestone is the old *lapis divinus*, made as follows:

> Potassium nitrate,
> Alum, } of each 1 part.
> Copper sulphate,
>
> Fuse together, and add camphor $\frac{1}{60}$ of the whole.
> To be run into moulds and kept in a stopped bottle.

LEAD ACETATE is of use in conjunctival and lachrymal diseases. It should not be employed when the cornea is involved, else staining of the latter may result. A common formula is:

> Liq. plumbi subacetatis............ f. ℨj.
> Aqua destill...................... Oj.

Twenty per cent. of alcohol added makes a cooling external application to inflamed lids.

NITRATE OF SILVER is one of the best astringent caustics we possess, and it is found to act admirably in many external affections of the eye. It should never be used by the patient stronger than a half per cent. solution in distilled water. More powerful mixtures had better be applied by the oculist himself. Although the action of this silver salt is limited by the formation with the tissues of an inert albuminate of silver, it is always advisable to have at hand a solution of salt and some water. When strong solutions are used, the excess of the nitrate may be neutralized and prevented from affecting the surrounding parts by brushing the latter with the saline solution, which, with the resulting chloride and some shreds of albuminate of silver, can be removed by subsequent applications of pure water. This is the best method of applying solutions of silver nitrate to young children — as, for instance, in ophthalmia neonatorum (M. 34). If this remedy be persisted in for weeks or months it may stain the conjunctiva. The solid salt ought not to be applied to the conjunctiva, although such intimate mixtures as the "*mitigated stick*" may be used without danger. The formula for the latter is:

>Nitrate of silver.................. 1 part.
>Nitrate of potash.................. 2 parts.
>
>Fuse and run into moulds.

All preparations of silver nitrate should be kept out of the light.

TANNIC ACID, and mixtures of it with alum, glycerine, and water, are popular with some ophthalmic surgeons. My own experience of it has not impressed me favorably. The so-called *glycerite of tannin* has been recommended for trachoma (M. 45).

BANDAGES for the eye are of two kinds. The first is used when it is desired to exclude all light, and for this purpose nothing is better than well-washed white flannel. When the purpose is to retain dressings in place, cheese-cloth or muslin is much to be preferred. When once removed from the eye, a fresh bandage *ought* to be used, the flannel being saved, the cheese-cloth thrown away.

BORATED OR SALICYLATED COTTON WOOL, kept clean in a tin box, makes the best dressing for most ophthalmic cases.

GOGGLES for excluding the light and for protecting the eyes from dust and wind are made of a colored glass front, whose sides are fine meshed wire. They are, when complete protection is desired, to be preferred to *coquilles* (dome-shaped glasses), or plain colored spectacles, although the latter have the advantage so far as appearance goes. Opticians keep these protective glasses in various tints of "smoke" and blue.

SHADES are made of cardboard for one or both eyes. The home-made article is objectionable, in that it is usually fashioned so that it touches the lids or eyelashes. It then shortly becomes, surgically, unclean from the ocular secretions, and for this reason is a source of danger.

EYE CUPS are made to fit the edges of the orbit, brow, and nose, with the idea of effectually bathing the eye. Filled with the medicated solution, the cup is accurately applied to the orbital margin, the head is thrown back, and when in that position, the lids are frequently opened and shut until every part is thoroughly reached. Heat and cold can also be applied in this way.

FOR IRRIGATING THE EYE before and after operations recourse is had to the apparatus commonly employed for such purpose.

In applying the CAUTERY, the electric form is to be preferred to either the actual or that of Paquelin, for many reasons. The cautery "point" should be a delicate one, and heated white hot.

The best SPONGES are small "dabs" of borated cotton, which are thrown away after being used.

OPERATION PRELIMINARIES.

The patient's person and clothing should be clean. The conjunctival sac is well irrigated with boiled boric acid solution, and the eyelashes, lids, eyebrows, and cheeks thoroughly scrubbed first with a saturated solution of boric acid, and then bathed with a 1:5000 solution of corrosive sublimate. It is assumed that the instruments have all been placed in a 95 per cent. solution of carbolic acid for a few minutes, and, just before using, are transferred to a boiled saturated solution of boracic acid, or to boiled water alone. The surgeon's hands should be well washed with hot water and soap, and then disinfected. It is superfluous to add that the operator should not be less clean than the patient. A well-lighted (aseptic) room is chosen for the operation. The patient reclines upon an operating or other low table, with the head slightly raised and steadied. The surgeon stands in front or behind, as he wishes, and his assistant stands conveniently by in charge of the instruments.

LESSON III.

DISEASES OF THE EYELIDS AND CONJUNCTIVA.

Atropine Irritation — Pterygium — Pinguecula — Lithiasis — Blepharitis—A "Black Eye"—Herpes of the Lid—Stye or Hordeolum—Chalazion or Cyst of the Lid—Lupus and Epithelioma—Xanthelasma—Eutropion and Ectropion—Surgical and other Treatment of these Conditions —Hotz's Operation—Ptosis—Lagophthalmos—Symblepharon.

The more important diseases of the conjunctiva have been exhaustively treated in Mittendorf's work (see Preface) on the subject. It remains for me to speak of some additional affections of that membrane.

ATROPINE IRRITATION AND CONJUNCTIVITIS.

Atropine and its salts are now so extensively used that it is important to recognize a not uncommon idiosyncrasy which some patients, especially old people, exhibit. It sometimes happens that even after one or two instillations of a weak solution the conjunctiva becomes vascular and thickened; a muco-purulent discharge is set up and all the evidences of an acute inflammation show themselves. At the same time the skin of the lids appears puffy, shiny, excoriated and reddened. These symptoms, due as Treacher Collins has shown, to the local irritant effects of the drug, disappear if the atropine be stopped and zinc ointment be applied to the lids. In such cases also duboisia (23) should be substituted for the atropine and boracic lotion (26) used as a collyrium.

PTERYGIUM. This is a fleshy, triangular, hypertrophy of the conjunctiva with its apex applied to the cornea and its base towards one of the canthi. The origin of the thickened growth is a curious one. A marginal ulcer of the cornea (69) forms, and in healing incarcerates a minute portion of the ocular conjunctiva. This throws the latter into a triangular fold which later on enlarges, attaches itself to the corneal tissue, probably by proliferations of its cellular constituents, and advances towards the centre of the cornea, which it sometimes reaches.

Treatment. Excision is the only treatment that accomplishes anything. The corneal attachment of the growth should be carefully and evenly dissected away from its bed, care being taken to avoid injury to the deeper tissues. The body of the pterygium is next excised in its entirety. The sound conjunctiva having been undermined is stretched over the vacant triangle and joined by sutures.

PINGUECULA. This is a small yellowish elevation on the conjunctiva occurring commonly within the inter-palpebral slit and usually on the nasal side. It is an inoffensive growth, is composed of connective tissue, seldom attains a large size and is probably the result of irritation from foreign bodies. Its removal by means of forceps and scissors may be undertaken, if considered desirable.

LITHIASIS. Chalky degeneration of the meibomian secretions (5) may often be noticed as white spots on the conjunctiva about as large as a pin's head. They are sometimes surrounded by a zone of injected blood vessels, and may be a source of considerable irritation. If productive of symptoms they should be removed under cocaine by first making a small incision over them and then turning out the calcareous particles with a needle.

BLEPHARITIS MARGINALIS, called also *Tinea tarsi*, may or may not be a true eczema of the border of the lids. It is a very chronic affection, lasting often for years, and is frequently accompanied by chronic conjunctivitis. In such cases both affections should be treated together (M. 71). The chief sign of the disease is the formation of crusts or scales along the lid margin. These when removed expose a glazed, reddened or moist surface. The small crusts, which should not be mistaken for eggs of pediculi sometimes laid in this situation, adhere to the base of the cilia which often become stunted and broken. After a time the disease affects the root-sheath of the cilium, the bulb atrophies and the lids become more or less destitute of lashes. The *symptoms* are not, as a rule, urgent, but a feeling of irritation and heat in the eye, which is always aggravated by exposure to wind and sun, is usually noticed. After a time the eyes, having lost their hairy defenders, suffer from the entrance of dust and other foreign bodies.

Treatment should first of all be directed towards removal of the crusts. Very few patients persevere in this endeavor as they should, and it is accordingly

FIG. 1.

often a wise measure to remove with the forceps (Fig. 1) every eyelash that harbors the scabby exudations. This prevents the re-formation of the crusts, and gives the remedies employed a better chance to reach the seat of the disease and set up healthy action in the parts affected. The best way to remove the crusts is to soak them well with a hot solution of sodic carbonate, a 2 per cent. solution diluted with its own weight of boiling water.

After the removal of *all* the scabs, a one-half per cent. mixture of the red (or yellow) oxide of mercury with cold cream should be thoroughly rubbed into the edges of the closed eyelids. This may be done in the evening, a few hours before retiring, while a boric acid lotion (26) should be applied several times during the day. When the case is one of eczema, with moist crusts, swelling of the lids, and conjunctivitis, various measures have been advocated. I have seen admirable results from an ointment recommended in cases of eczema by Dr. Zeisler of this city. The formula is:

    Resorcin 0.30
    Lac sulphur....................... 1.00
    Lanoline 5.00

It goes without saying that patients suffering from blepharitis should avoid dust, heat and wind, as much as possible. If necessary, they should wear protective glasses (30). They should not smoke themselves, nor allow their eyes to be irritated by remaining in a smoking-room. The general health is worth looking after; indeed, it may be that a strumous diathésis lies at the bottom of the disease.

It *often* happens that a blepharitis is perpetuated by "eye-strain" (R. 10). Proper glasses should in such instances be ordered, especially if there be any astigmatism (R. 42) present. Whatever the treatment may be, a complete cure is not, in the majority of cases, to be expected inside of several months.

ECCHYMOSIS of the lids, with its usual accompaniment of subconjunctival hæmorrhage, constitutes what is popularly termed a "black eye." Where a definite blood-clot has formed within the palpebral tissues, the common practice of incising the skin and allowing the blood to escape is a good one. Antiseptic dressing should be subsequently applied. The average chemosed eye will be best treated with an evaporating lead lotion (28). Unless treatment is resorted to within two days, no remedy will be of use.

It is then best to cover up the discoloration with flesh-colored paint. It will pay every practitioner to keep some water colors for the purpose. No production of his artistic hand will be more appreciated than that which disguises such a noticeable blemish.

HERPES OF THE LIDS resembles herpes zoster elsewhere. It is not of frequent occurrence, but ought to be easily recognized. The herpetic vesicles are disposed about one or more cutaneous branches of the fifth nerve, and the pain accompanying the disease is severe and of a neuralgic character. It may also attack the cornea, conjunctiva, and even iris, and where it does so the results may be serious. Morphia should be given to relieve pain, quinine for its specific effect, while such local applications as hot belladonna fomentations are useful.

When the cornea is affected—as in a case recently seen by me—a mixture of eserine and cocaine (24) acts very well, both in lessening the pain and subduing the corneal inflammation.

STYE OR HORDEOLUM. This is a very common lid affection and may be regarded as a palpebral "boil." When it occurs near the outer canthus the œdema of the lid or lids is often considerable. This is probably due (Lang) to the blocking of the lymphatic stream which empties into the larger channel near the ear. In children there is usually a good deal of pain and sometimes fever. In its earliest stage, before pus has formed, the stye may sometimes be aborted by pulling out the eyelash which runs through it and painting the tumor with strong tincture of iodine. If this fails the point of a Beer's knife—or some similar instrument—should be pushed into the centre of the tumor and its contents evacuated. Subsequently a poultice may be applied and then, in a day or so, a mild mercuric ointment (26) should be rubbed over the diseased part. Hordeola, like boils elsewhere, are liable to recur and when they do careful search should be made to detect some error of refraction (R. 22), some constitutional cause, or impropriety of diet, likely to account for such a state of things. Frequently, as in anæmic girls, a course of iron and fresh air is what is chiefly needed.

EYELASH IN A PUNCTUM. Careful inspection of both puncta (4) as a routine observance would prevent one's overlooking this little accident, but the possibility of its happening should always be borne in mind. Until removed it creates a good deal of disturbance as well actual conjunctivitis.

CHALAZION. Cyst of the lid. This is a small, painless, hard, slow-growing and slightly movable tumor imbedded in the tarsal cartilage. It is a "retention cyst," being generally produced by the obstruction of a meibomian (5) tubule. The contents are usually cheesy, but the tumor sometimes resembles a fibroma in hardness. Local applications do little or no good. As the tumor generally "points" towards the conjunctival surface of the lid it is best to evert the latter and empty the cyst in that direction. Some surgeons use a special clamp for this purpose (see Fig. 2.)

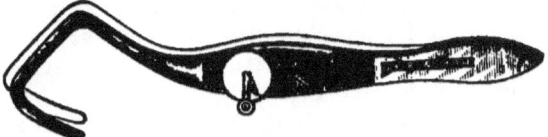

FIG. 2.—SNELLEN'S LID CLAMP.

In all cases the parts should be anæsthetized by means of a four per-cent. solution of cocaine. The clamp having been applied, the lid is everted, an incision made the whole length of the tumor (whose position is indicated by a purplish discoloration of the mucous membrane), and the cyst contents evacuated by means of a small scoop or the end of a director. In large chalazia it is well to arm a probe with a little cotton wool, and brush out the cavity with a drop of strong nitrate of silver solution (4 or 5 per cent.). This will effectually prevent their return. The patient should always be warned of the fact that after an operation for the removal of the contents of a meibomian cyst the vacuum is filled with a blood clot and the tumor feels larger than ever. In a few days, however, absorption begins and the tumor gradually disappears.

Tarsal cysts may be produced by eye strain (R. 10) blepharitis (34), and other diseases, local and general. When they occur and are multiple this fact should be borne in mind.

Rodent ulcers, epithelioma and lupus affect the lids as elsewhere. In all three diseases prompt and early treatment by the actual cautery or other caustics will prove effectual. Later on a plastic operation will be necessary.

Xanthelasma is a sufficiently common affection of the lids and is almost exclusively confined to the female sex—especially those, as Meyer says—who suffer from migraine and affections of the liver. It forms an irregular yellowish patch or patches and is made up chiefly of fibrous tissue—not fat. It is a harmless growth, but may be removed if the patient objects to its presence.

Entropion.—This is a term which indicates a turning in of the edge of the lid. It may result from contraction of the orbicularis in old people with flabby lids, from burns; or from other injuries, but is almost always caused by granular lids (M. 45). This last disease first produces scarring (and contraction) of the mucous membrane, and later on, irregular atrophy and consequent incurvation of the tarsus. The eyelashes are secondarily affected, and some of them turn down (4), touch and irritate the cornea. If there be two regular rows of cilia, one is very apt, for a while at least, to retain its outward curve, while the other curves in upon the sensitive globe. Such a condition of affairs is termed distichiasis. When the curvature of the lashes is irregular, or if but a few of them are thus affected, the term trichiasis is employed to describe it. In most of these cases the situation is made worse by a shortening of the interpalpebral aperture; the lids are too closely applied to the eyeball; the patient cannot separate them widely enough, and all the symptoms are aggravated thereby. This may be remedied by a simple operation called CANTHOPLASTY, and it may be performed alone or as an adjunct to other operations upon the lids or lashes. The external commisure is cut through in the horizontal line and directly outwards with a pair of straight scissors. The sharp-pointed blade of the latter is entered underneath the lids at the outer canthus, and the skin wound made a few millimetres longer that in the mucous membrane. The subcutaneous injection of a few drops of cocaine (4 per cent. solution) will render this a comparatively painless operation. The conjunctival edges are now well separated from the underlying tissues, and with three stitches are evenly joined to the margin of the skin wound. An antiseptic lotion, and the removal of the stitches, in from four days to a week, complete the cure. If the entropion be well marked, an efficient operation constitutes the best means of remedying *all* the evils attendant upon that condition, but in many cases it will suffice to do a canthoplasty and destroy a few troublesome cilia. Do not, however, temporize by pulling them out unless the patient positively refuses to undergo an operation. In that case remember it is the fine, short, and colorless hairs, that do most mischief, and are just the ones most likely to be overlooked, unless one have sharp eyes or use a lens (9).

EPILATION is practiced with special cilia forceps having broad and smooth ends (Fig. 1.).

It is best to destroy the incurving hairs of trichiasis, if they are not too numerous, by one or both of the following measures: 1st. Michel's method, electrolysis. A platinum needle is connected with the negative pole of a 20-volt battery, and plunged accurately into the ciliary follicle, the positive pole being grasped by the patient. The action of the battery, as evidenced by the disengagement of hydrogen from the neighborhood of the follicle, should be kept up for 60 seconds. Even with a previous injection of cocaine this is a painful procedure. 2d. Snellen's method. Thread a small and sharp needle with both ends of a fine silk thread. Enter the former at the base of a cilium, push it underneath the palpebral skin, and bring it out six or eight mm. from the lid margin. As the doubled thread is drawn through, ensnare the lash and draw it bodily into the substance of the lid.

Hotz's operation for entropion.—When from one cause or another (usually resulting from long standing trachoma or granular lids) the preceding measures are found or judged to be inadequate for the cure of the entropion and triachiasis, a more radical operation is necessary. A volume might be devoted to a description all of those that have been from time to time devised. Taken all in all, that of Hotz is to be preferred. He aims to make the lower edge of the lid wound adhere to the upper edge of the tarsus and so by a sort of leverage action draw the palpebral border with its incurved cilia outward.

Ether or chloroform is given and a lid spatula may or may not be used. An assistant now draws up and holds the skin of the (upper) lid firmly against the brow while the operator puts it on the stretch with forceps in an opposite direction. An incision is made horizontally from a point 2 min. above the inner canthus to a corresponding point above the outer commissure. If the skin be now left free this incision will be found to be *curved* and to correspond with the superior edge of the tarsus. The lower edge of the wound is now drawn down with forceps and the surgeon dissects some of the fibres of the orbicularis from the upper third of the tarsus. The bleeding having stopped, three or four sutures are inserted by a curved needle, first into the upper edge of the wound, then through the upper edge of the tarsus and some of the tarso-orbital fascia just above it, and finally through the lower margin of the skin wound. The bleeding having stopped and the wound cleared of clots and well irrigated, the ends of the thread are firmly tied together binding both edges of the incision to the upper margin of the tarsal plate. Iodoform or other antiseptic dressing is now applied and the stitches should be removed *in two days* or on the approach of suppuration.

ECTROPIUM. In old people when a portion of the musculus orbicularis becomes atrophied, the lower lid is especially prone to resign its close application to the globe and sinks down, carrying with it the punctum lacrymale. As a consequence the tears flow over the cheek and produce excoriation of the skin and edge of the lid. This in its turn brings on œdema of the parts, chronic conjunctivitis and finally spasm of the remaining orbicularis fibres so that the lid is everted. This particular form of the disease is terned *senile ectropium*. Scars from burns and wounds may also bring about the same condition, but the great majority of examples of entropium are the result of muscular spasm caused by œdema of the palpebral conjunctiva.

The Treatment should be directed first towards the removal of the cause of the trouble. The excision of cicatricial tissue with transplantation of skin will afford ample opportunity for the exercise of the surgeon's best skill, and the rules to be observed do not differ from those in vogue in other skin and mucous membrane regions elsewhere. Do not forget that pieces of skin have been removed *en masse* from the arm and other situations and have taken kindly to their new position in the facial region. The value or necessity of the pedicle has probably been overrated hitherto. For muscular entropium many operative measures have been employed. One of the best of these—easy to perform—is that of Snellen. A double-needled silk thread is used, one needle being entered where the everted conjunctiva is most prominent and brought out through the skin 2 cm. below the lower lid margin. The other needle is passed in the same way, the points of exit in the skin being about 1 cm. apart. Traction is now made upon the threads, the mucous membrane is pulled *down* while the palpebral margin is assisted into place. The ends of the thread are tied over a piece of rubber, to protect the skin. In the meantime any œdema, conjunctivitis, or other lid affection should be treated, in the hope that the patient will be able to get along without the sutures—two or more of which may be required.

In spite of this and similar devices a cutting operation may be needed. That recommended by Adams is a good one, but to be successful and not to leave a deficiency at the border of the lid or an ugly scar in the skin very accurate coaptation of the edges of the wound should be secured.

A piece of the lid in its whole thickness is removed with the mucous membrane, as pictured in Fig. 3. The edges are then carefully brought together and dressed.

PTOSIS.—Drooping of the upper lid may result from a number of causes. Of these the commonest is paralysis of the third nerve (137), which supplies the levator palpebræ. Next in order of causation come thickening and increased weight of the lid from diseases (chronic inflammation, trachoma, etc.), congenital deficiency of the elevator muscle, and (when it is bilateral) finally wounds and adhesions. Apart from the deformity, which is very noticeable, the falling of the lid over the pupil directly interferes with vision.

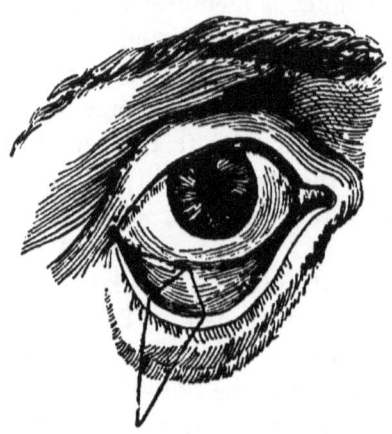

FIG. 3.

Treatment. —After electricity and other appropriate treatment have failed, this condition is remedied by an operation. The usual method is to excise a horizontal oval-shaped piece of skin from the lid. To remove enough to allow of vision, and yet not so much as to prevent closure of the eyè during sleep, is the problem. A preliminary observation should be made by pinching up the palbebral skin with a pair of forceps, and thus judging of the amount to be removed. Subsequently the edges of the wound are brought together by sutures.

LAGOPHTHALMOS.—This is the condition opposed to ptosis, wherein, from paralysis of the orbicularis, the patient is unable to close the eyes. Literally translated, it means "hare eye," as that animal is said to sleep with its eyes open. The diagnosis is easily made if the patient be told to shut his eyes. When the disease is one-sided, as it commonly is, these efforts are productive of a curious result, viz.: the eye of the affected side assumes the position it occupies during sleep, and is plainly seen to roll up under the unclosed lid. This is a fortunate position, as it serves to protect the cornea from the dessicating effects of the atmosphere, and from injury by foreign bodies. Lagophthalmos is almost always produced by paralysis of the portio dura, but it may also result from anything which brings about undue projection of the eyeball, as staphyloma corneæ (76), extreme myopia, Graves' disease (122), orbital (121) and intra-ocular growths.

Treatment.—In paralytic cases, and in some of those due to the other causes mentioned, a simple operation, termed *tarsoraphy*, will be useful. By means of it the interpalpebral slit is both shortened and narrowed.

A short strip of skin, the length of which will be determined by the effect desired, is removed from the margins of the lids at the outer canthus. The cilia with their bulbs are included in the excision, and the pared edges are sewed together. This operation will not be undertaken until the effects of remedies directed towards the removal of the cause of the lagophthalmos have been tried.

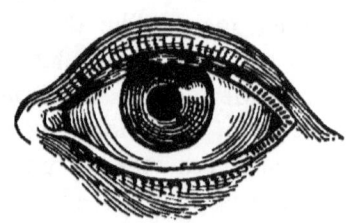

Fig. 4.

SYMBLEPHARON.—This is the term applied to the abnormal union of the ocular with the palpebral conjunctiva. It may be partial or total, and is most frequently produced by escharotics, such as lime, molten lead, acids, or strong alkalies, introduced into the conjunctival sac. Adhesion does not occur for several days or weeks after the accident, and it is extremely difficult to prevent it.

Treatment.—In slight cases the cicatricial bands are to be cut through, and the edges of the healthy conjunctiva united over the raw surface by sutures. In more extensive cases, after separation of the conjunctival surfaces, one or both wounds may be covered by conjunctival flaps or with mucous membrane transferred from the lip or from the rabbit's conjunctiva.

LESSON IV.

DISEASES OF THE LACHRYMAL APPARATUS.

Épiphora—Lachrymation—Dacryocystitis—Lachrymal Abscess—Slitting Up the Canaliculus—Passing the Nasal Probe—Treatment of Lachrymal Obstruction by the Syringe of Meyer or Anel.

EPIPHORA is the name usually employed to designate a flow of tears over the lower lid margin. LACHRYMATION refers more particularly to an increase in the supply of tears. Anything which interferes with the drainage (5) of the external eye will produce epiphora; any cause which stimulates the formation of tears produces lachrymation. Among other causes of epiphora, besides lachrymation, may be mentioned: (1) Those that produce misplacement of the puncta (4), such as surgical and other wounds in the neighborhood, paralysis of the facial nerve (by which the orbicularis loses its tone), and laxness of the palpebral tissues dependent upon senile changes; (2) Obstruction at some point in the lachrymal canal or nasal duct. The latter is the commonest and most important cause of epiphora, which will hereafter be considered merely as a symptom of this class of diseases.

DACRYOCYSTITIS.—There is a certain resemblance between inflammation of the urethra and the same affection of the tear passages. Both usually start from infection supplied from without—in the case of the latter from chronic maladies of the lids, such as blepharitis (34), trachoma (M. 45), diseases of the nose and nasal duct; both tend to the formation of pus and to stricture of the canal; and both have, generally speaking, the same treatment. When the mucous lining of the canaliculus and tear sac becomes inflamed, it is called dacryocystitis. The acute disease shortly passes into the chronic stage, or it is chronic from the beginning, and we have as symptoms of the latter mainly epiphora, increased by wind and sun, and the hypersecretion of mucus mixed with some pus. For a time these abnormal secretions are carried along the nasal duct into the nose, but after a while the whole mucous membrane of the canaliculus, sac, and nasal duct becomes involved, distension of the sac takes place, and the muco-pus regurgitates through one or both puncta into the conjunctival sac, and may be seen as flocculi floating in the tear drops. In all such cases, firm pressure (5) should be made over the region of the sac at the inner canthus. Even before the internal enlargement becomes so pronounced as to show itself as a decided swelling in such region, this procedure will cause the muco-purulent fluid to issue from one or both puncta. This settles the point, and proves at least the existence of an obstruction in or below the lachrymal sac.

The *obstruction* may be due merely to swelling of the chronically inflamed mucous membrane, or it may mean an organic stricture closely resembling an urethral stricture. The practitioner will find the following a useful guide to diagnosis in these cases: *When the sac is enlarged, and pus can be squeezed out of the puncta, an organic obstruction (stricture) is present; but if there be little or no cystic swelling and the secretion is mostly mucus, the obstruction is a swollen mucous membrane.*

In the first instance the introduction of probes, combined with slitting up one of the canaliculi, will be the only effective treatment. In the latter instance— in the so-called *mucocele*—washing out the lachrymal tract (55) through one of the puncta will be efficacious.

LACHRYMAL ABSCESS.—A distended sac in these chronic cases is always liable to inflame and form an abscess. When that occurs, an erysipelatous blush surrounds the seat of the disease, there is considerable swelling of the lid and sometimes of the face, while the pain may be severe. If left to itself, the pus points underneath the skin about the sac, the abscess opens and leaves an ugly fistula, which is often difficult to heal.

Treatment. If seen early either the lower or the upper canaliculus should be slit up with a knife used for the purpose, and a probe passed through the stricture into the nose. This is usually sufficient to stay the further progress of the disease. If seen late, when the skin over the abscess is very thin, it is better to open the latter, reduce the swelling by appropriate applications and attend to the stricture subsequently. The scar left is insignificant.

Fig. 5.

SLITTING UP THE CANALICULUS is performed with the knife (Fig. 5) before referred to, as follows: If ambidextrous the surgeon always sits in front of his patient. If he wishes to use his right hand in every case he will stand behind the patient's head when he operates on the right eye; the left canaliculi are more easily reached from the front. To open the lower canaliculus—the usual one—the punctum is everted and the point of the knife, edge upward, entered at right angles to the lid margin. The palpebral skin is now drawn towards the outer canthus with the disengaged hand and the handle of the knife depressed until it is almost horizontal. It is now pushed towards the sac until its nasal wall is felt. Keeping its point steadily in that position the handle of the knife is partially rotated so that the edge of the blade now looks upwards and slightly inwards. The handle is now carried up and slightly past the median line, cutting through the wall of the canal within the lid margin, and is again rotated to be used as a cutting probe, and passed down into the nasal duct as described below. An anæsthetic is not usually necessary, and bleeding from the nose should follow the operation, showing that an open communication now exists between the conjunctival sac and the meatus. Lachrymal probes may be passed after 24 hours.

PASSING THE PROBE. Stricture of the nasal duct is as difficult to *cure* as is the urethral stricture and it may be necessary to pass probes (bougies) two or three times a week or oftener, for several months. I am

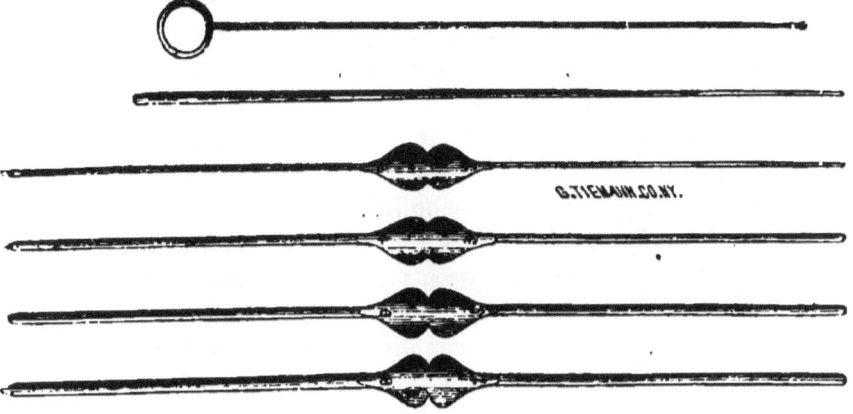

FIG. 6.

greatly in favor of teaching the patient to do this himself and to instruct him to keep it up at increasing intervals for a year or more. The difficulty is that the sufferer getting *relief* from an operation and the subsequent half dozen probings administered by the surgeon, gives up treatment and the disease relapses. The probes used are of all sizes and shapes. Those devised by Bowman, made of silver, (Fig. 6) and numbered from 1 to 8 answer most purposes. Passed along the opened canaliculus to the posterior wall of the sac, with the lid margin kept on the stretch, the probe is elevated and carried slightly past the median line until it, while almost touching the brow, points downwards, outwards, and slightly backwards. It should now pass, *without employing undue force*, into the nose. As large a size as possible should be used and the probing should be done daily, while the patient should take lessons in passing the instrument himself. Care must always be observed not to make a false passage. If persevered in, a cure of lacrymal obstruction when not due to diseases of the bone (osteomata, syphilis and the like) can be confidently expected. With the observance of all precautions a fine probe may, in the manner that the canaliculus knife is entered (52), be passed through the punctum along the canaliculus and into the nasal duct. This was Becker's plan and in cases of mucocele or slight stricture it has much to recommend it.

FIG. 7.

Anel's or Meyer's syringe. A canaliculus once-opened in the manner above described seldom or never closes, so that the natural drainage-function of the parts is destroyed. The operation should not on that account be lightly undertaken. For most cases of mucocele—as before stated—systematic syringing of the passage from punctum to nasal meatus is sufficient in many instances to bring about a cure without having recourse to probing or the cutting operation. In any case, however, treatment of the lining membrane of the tear passage is indicated. For this purpose boric acid lotion, with the addition of zinc sulphate, 1 grain to the fluidounce, makes a very good astringent application and is much used. Some surgeons prefer oily preparations—such a prescription as this, for example:

    Menthol................ .... 0.50
    Benzoinol 10.00

Whatever be the remedy, a syringe such as is shown in Fig. 7 is used. The sac contents are first squeezed out, the point of the syringe entered at the punctum, passed along to the sac, as in probing (53), directed downward and the nasal duct flushed into the nose. If the punctum be too small for the entrance of the syringe point, it must first be dilated with a fine Bowmans' probe. Cocaine may be applied as a preliminary to this procedure, but the pain is inconsiderable. In all such cases any accompanying conjunctival or nasal disease should by no means be neglected. Lang advises that, as in the treatment by probing, the patient be taught to continue the lacrymal "flushings" at home, with a fine point attached to a rubber tube and bulb.

LESSON V.

DISEASES OF THE CORNEA AND SCLEROTIC.

Arcus Senilis—Keratitis—Phlyctenular Keratitis—Foreign Bodies in the Cornea—Interstitial Keratitis—Punctate Keratitis—Ulcers of the Cornea—Spreading and Nonspreading Ulcers—Hypopyon and Onyx—Paracentesis of the Anterior Chamber—The Use of the Cautery—Senile Ulcer—Opacities of the Cornea—Nebula, Macula and Leucoma—Staphyloma Anterius—Tatooing the Cornea—Scleritis and Episcleritis

ARCUS SENILIS.—The cornea retains its central transparency in a wonderful way until quite old age and it rarely happens, except as the result of injury or inflammation, that vision is interfered with in consequence of degeneration of its tissues. Marginal changes are not uncommon. The most frequent of these is the so-called arcus senilis, although it is not necessarily a sign of senility. It presents itself in the form of a narrow grayish crescent, placed above or below, or it may extend entirely round the cornea. It is composed of corneal cells which have undergone a true fatty metamorphosis.

Inflammation of the cornea is termed KERATITIS. The inflammatory process may affect the external epithelium and superficial layers or it may extend as in ulcer (70), and parenchymatous keratitis (66), to the true tissue of the cornea. Finally, it may be confined chiefly to the endothelium—the membrane of Descemet (69).

PHLYCTENULAR KERATITIS.—This is substantially the same disease that one finds in the conjunctiva [M. 3.] and the little phlyctenulæ or ulcers which characterize it are sometimes seen affecting both cornea and conjunctiva.

In this little colored child* we have a good example of what is commonly known as phlyctenules of the cornea. Her mother, who comes with her, gives the following account of the case:

Three weeks ago the left eye began to water a little, and the child complained that bright light hurt it. Simultaneously with the discharge from the eye the nose began to run. Both these symptoms got gradually worse until about ten days after they were first noticed; the right eye and the right nostril also became affected, and began to discharge a watery fluid. The child henceforth kept herself shut up in a dark room or curled herself up in a corner, fearful lest the light should get to her eyes. Her appetite began to fail, she took no part in her usual amusements, and sometimes complained of pain in the eyes. Just now, as we examine her critically, we find that the child has her face buried in the angle formed by her bent arm, which she leans upon her chair. Disengaging her face, we find the eyelids somewhat

---

* Clinical Studies of the Eye. The author's paper in the *North American Practitioner* for Sept., 1890.

shiny and swollen. Both they and the cheeks are dotted over with moist eczematous patches. The upper lip is swollen, eczematous, and covered with nasal discharge. Pursuing our investigation still further, we find the moist crusts of acute eczema behind the right ear, and there is a thin discharge from the external meatus. The left ear is unaffected. The ear trouble has been there, the mother says, for over a month. The child now resists any attempt to open the lids, which are kept tightly closed. She will show us her tongue, which has a white coat. Formerly the term *photophobia* was believed to give a proper description of the condition present, but it is easy to show that there is no true fear of the light. We shall put a few drops of a 3 per cent. solution of cocaine into the little one's eye, and although the effect, so far as the retina is concerned, will be rather to irritate the latter (by dilating the pupil and admitting more light to the eye), still a decided amelioration of the supposed photophobia soon takes place. I now do this, putting, at short intervals, several drops into both eyes, and while we are waiting for it to produce its characteristic effects, we will proceed to discuss this and other matters connected with the disease. The cocaine relieves the photophobia (so-called) by numbing the terminal filaments of the fifth nerve—the sensitive nerve of the cornea. Iwanhoff thinks the corneal distress is caused by the irritation of the delicate nerve filaments, produced by wandering leucocytes as they enter the corneal tissue at the limbus and travel forward to form nests of cells immediately underneath the superficial epithelial layer. These round cells in their passage

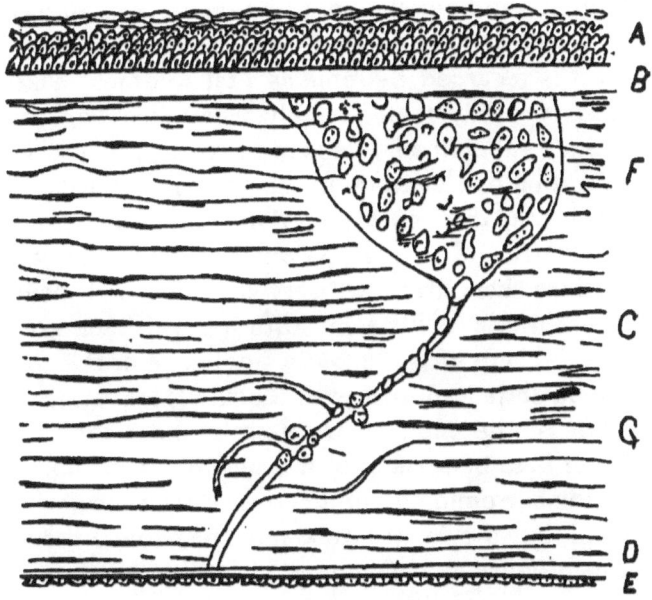

FIG. 8.

A. Anterior Epithelium.
B. Bowman's Membrane.
C. Substantia Propria.
D. Descemet's Membrane.
E. Endothelium.
F. Phlyctenule.
G. Nerve Filament pressed upon by emigrant cells.

irritate the minute nerve branches which ramify throughout the cornea, and give rise to reflex spasms of the orbicularis palpebrarum. The accompanying diagram (Iwanhoff) will explain this pretty theory, and help to give one an idea of the pathological histology of this disease and indicate what a phlyctenule really is. Phlyctenular keratitis occurs mostly in children out of health—often in strumous children—and very often, as we have seen to-day, in those who are subject to attacks of acute eczema. In fact it is regarded by some as a true corneal eczema. At first the phlyctenula merely shows (see diagram) as a minute subepithelial deposit of round cells—a slight infiltration or elevation on the corneal surface. This soon breaks through the overlying epithelium, and we have a true ulcer. These vesicles, as well as the ulcerations themselves, are very small, but well defined. One will see a little pit with yellowish or grayish sides. There may be but a single one or the cornea may be dotted all over with them. More or less injection of the small ciliary vessels about the corneal border is to be seen, and in some instances not only are the surrounding conjunctival vessels similarly enlarged, but they may extend into the corneal tissue, so as to produce a true pannus very like what one occasionally sees with granular lids. As a rule the disease improves rapidly under treatment, and in a few weeks the small patient is well. This is not always the case, however. The ulcerations may become so extensive as to be serious, and may even go on, in unhealthy children not properly fed, to perforation of the cornea and to destruction of the eye. In nearly all instances cicatricial opacities remain. If the ulcers have not been deep, and efficient treatment has been resorted to, these scars finally disappear.

This, by the way, is true of all superficial corneal scars as we find them in children. One should always feel encouraged to persevere (by massage with mild yellow or red mercurial ointment, ½ to 2 per cent., etc.) in attempts to remove these damaging interferences with vision.

Turning to the small patient who forms the text for this ophthalmic sermon, we find, as anticipated, that the cocaine has greatly relieved the reflex spasm of the lids, so that we are able, by the exercise of a little tact, to get a good view of both corneæ. The small, discrete, grayish ulcers—one in the left, two in the right eye, all of them near the centre of the cornea—are plainly seen against the dark-brown background of the iris, and we notice that the circumcorneal injection is considerable. In a previous lesson on "Iritis" I remarked that in young children well-marked "photophobia" means, in nine cases out of ten, phlyctenular keratitis, but one should never be contented with a knowledge of this fact alone. To see the phlyctenules, to observe their size and situation, to count their number, and to decide what stage of development they have reached, is the only common-sense method of diagnosis. We must observe, as we do in this case, whether there is any concomitant disease of the other ocular structures; that there is no iritis and no conjunctivitis. Practically speaking, iritic complications are rare, while phlyctenules of the conjunctiva (especially at the sclero-corneal margin) are not uncommon and should always be looked for. There are none in this case.

Now, how should we treat this little patient, and, incidentally, what is the treatment of phlyctenular keratitis? The child before us has that form of the disease known as *solitary phlyctenulæ*. I do not know that therapeutically this classification is of value, but it has a clinical significance, for it serves to distinguish those cases where there is no tendency towards the formation of more than one or two ulcers from others in which the cornea becomes very shortly the seat of numerous phlyctenules.

The first thing to do, in every instance, is to look well after the general health. The daily life of the small patient should be carefully inquired into, and one should insist upon its being compelled to conform as closely as possible to the ordinary laws of hygiene. These are not mere empty words, such as is the fashion nowadays to employ in embellishing papers on therapeutics An over-, under-, or badly-fed and housed child will have a flourishing crop of pustulous ulcers for an indefinite number of weeks or months, in spite of other treatment, while these will go on to reparation at once if a programme of living which includes common-sense diet, fresh air, and regular bathing, be adhered to in addition to the other remedies. Further, the state of the bowels (constipation or diarrhœa is often present), blood (anæmia, chlorosis, scrofula, etc., are to be looked for), and skin (eczematous eruptions are frequent) should not be neglected.

Notwithstanding the apparent photophobia, do not allow the child to remain curled up in a dark corner all day, but see that he or she is taken out often for fresh air and sunshine.

Whether the disease be a true eczema or not, I believe in giving Fowler's Solution (♍ j, to be increased gradually) three times a day after food, and I think benefit will be derived from it, more particularly in those instances where general eczema is present. For any accompanying facial eczema, also, I have always seen the best results from applying the same ointment to the face that I would apply to the eye, viz., a ½ to 1 per cent. mixture in "cold cream" of the yellow oxide of mercury. This mixture I give to the patient for use at home (a piece the size of a pinhead to be put into the eye two or three times a day), and if necessary use a stronger mixture (2 to 3 per cent.) myself once or twice a week. In the use of both these ointments I am guided by the amount of irritation which they cause. If the eye becomes "red" under the use of one or the other, I stop them for a time or change the treatment. Sometimes I have found calomel of use,, instead of the weaker ointment. This should be dusted into the eye with a camel's-hair brush once or twice daily. The only objection to it is that it requires a skilled hand to do the dusting effectually and properly. In all instances I have been in the habit of using atropine to dilate the pupil; once a day is, as a rule, often enough to apply it—a few drops of a one per cent. solution. If eserine be preferred (as it may be when the phlyctenulæ are numerous, when they are peripheral, or when they threaten to perforate the cornea), it is well to combine it with cocaine. A good formula is:

℞ Eserinæ sulph...................... gr. i.
    Cocainæ mur....................... gr. vj.
    Liq. hydrarg. bichlor., 1 : 5000....... fl. ℨ j.

M. et sig.—Two drops to be put into the eye three or four times a day.

I would still keep the pupil dilated with atropine, as it probably does not interfere with the curative effects of the eserine.

The relief of the "photophobia," or spasm of the lids, is important. Spraying the eyes with ice-cold water, or dropping it on the closed lids, as Oppenheimer suggests, in conjunction with the occasional instillation of cocaine (2 per cent. solution), will be found to give great relief and be very grateful to the patient.

The child should wear a shade in the house, and a light porous bandage over the affected eyes when he is taken out. Later on, smoked or blue glasses (*coquilles*) (30) will answer both purposes.

To keep the eyes perfectly clean with some antiseptic solution, which shall also act upon the ulcerations, is good surgery. I have never been able to see the necessity, however, for washing out the eyes *every hour*, as some authorities advise. It is entirely too frequent. Aside from the difficulty met with in getting the eyes open, it seems to me that the worry occasioned the child in attempting to accomplish it does more harm than good. On the other hand, a mild solution (3 per cent.) of boric acid, or of chlorine water, applied with a dropper four times a day and once during the night, is decidedly helpful. Some firmness and a little coaxing, joined to the effects of a few drops of the cocaine solution, will generally enable the nurse to apply the antiseptic bath.

Finally, do not forget that the patient is nearly always "below par," has a poor appetite, a coated tongue, and an indifferent digestion. My favorite prescription under these circumstances is: Tr. nucis vom., ℞ ij, in ℥ ij of "beef, iron and wine." Children take this mixture well. If necessary, peptonized milk should be given with the other food. All such indigestible trash as ice-cream, greasy cakes, candies, *et hoc genus omne*, should be proscribed.

FOREIGN BODIES IN THE CORNEA.—The "something" which gets into a patient's eye is usually washed towards the inner canthus, out of harm's way, by the stream of tears excited by its presence. But if it has sharp edges or corners it may stick into the cornea—if, indeed, the force that sends it into the eye has not already driven it into that structure. *Removal* of these is usually done under cocaine, by means of a "spud" (Fig. 9).

FIG. 9.

It is wise to try and pry the object out of its bed rather than to scrape it away. Iron or steel chips are readily attracted by the electro-magnet, and when they are deeply situated or project into the anterior chamber, this is the safest method of removal. To prevent infection (31) of the wound produced by the foreign body, a lotion of boracic acid with corrosive sublimate (27), should be used several times daily for a week.

INTERSTITIAL KERATITIS is a very chronic inflammation of the cornea, the result in most instances of inherited syphilis The child—for it is essentially a disease of childhood—nearly always presents some of the well-known signs of congenital syphilis:—the notched (Hutchinson) incisor teeth, the sallow skin, the anæmic lips, the broad depressed nose, the scars at the angle of the mouth, and the absent naso-labial depressions. The mother, if questioned, will be found to have had several miscarriages or dead-born children. Deafness, from internal ear disease, is a not infrequent accompaniment of this *parenchymatous* keratitis. It begins very insidiously as a slight grayish opacity near the upper margin of the cornea. At the same time the vessels in that region become injected and reach out to join the infiltration. The vascularity increases with the grayish opacity until a yellowish area is formed, to which the name " salmon patch" has been given. Little by little the opaque patch increases in size, other spots form and coalesce, until finally the whole cornea presents an irregular grayish-white aspect, intersected by numerous blood-vessels. The epithelium of the affected parts loses its smooth appearance and looks like ground glass. This stage of the disease may extend over several weeks or months. When the cornea has reached a condition of extreme opacity it remains unchanged for a longer or shorter period, and then begins to clear up. The process of absorption and repair now goes slowly on, translucent areas show themselves here and there, and finally portions of the iris and pupil can be seen, and after many weary weeks of waiting the cornea may become quite or almost as transparent as before. This is the usual course of the disease, and the patient's friends can be encouraged to persevere with remedies, however unpromising the outlook may seem. There is always more or less pain, lacrymation and photophobia, and sometimes spasm of the lids. Iritis, as well as more deeply seated inflammation of the eye, may be present. If the eyeball is tender and the tension (9) is lessened, these complications may be suspected.

Abscess and ulceration of the cornea are seldom or never present, although some pus cells probably collect about the inflammatory foci. Usually both eyes are attacked—one after the other. It often happens that as one eye is getting well the other becomes affected—a fact to be borne in mind in delivering an opinion to the patient or his parents. Proper treatment may shorten the duration of this disease, but it often lasts six or eight months under the most attentive care. *Prognosis* is generally favorable, but in view of the possibility of incomplete clearing up of the cornea, as well as the chances of iritis, chloroiditis and other complications, it is best to be guarded in this particular.

Treatment should first of all be directed to the general condition. When syphilis or struma is present constitutional remedies will be called for. In the former case the syrup of the iodide of iron is invaluable. To this short and intermittent courses of mercurous (green) iodide may be added. The pupil should be kept dilated with weak atropine drops (2 grs. to the f ℥ j) if there be pain or tenderness on pressure. Tonics, cod-liver oil, a good diet and fresh air are always advisable whatever the cause of the disease. I think I have seen good results from massage with mild mercurial ointment (26) after the acute stage has passed. Two other remedies must not be forgotten: 1st, fomentations with water as hot as can be borne for an hour three times daily; and,

2nd, the application of a small blister (2x2) to the temple about once a fortnight. These constitute a sort of routine practice in which I have much faith and I think they may be used with great benefit in the large majority of cases.

PUNCTATE KERATITIS may be known by the formation of numerous dots of lymph upon Descemet's membrane—the posterior surface of the cornea. These fine heads arrange themselves, as a triangle in the lower corneal semicircle with its base at the periphery and its apex pointing to the center of the cornea. This is not an independent disease of the cornea but is the outcome of serous iritis (129) and sympathetic ophthalmia (130).

Treatment.—This must be directed mainly to the disease with which the punctate spots are associated.

Hypodermic injections of pilocarpine (24) are of value.

ULCERS OF THE CORNEA. These form themselves into several clinical groups, but as commonly seen they may be divided into two classes, (1) *the simple, non-spreading ulcer;* (2) *the serpiginous or spreading variety.* Both lesions are usually the result of a wound of the cornea (however slight) with subsequent infection. For example, a grain of coal, a piece of metal, or other foreign body becomes embedded in, or some other agent inflicts a wound upon, the cornea of a patient who has a mucocele (51), blepharitis (34), or it may be some form of conjunctivitis. The micro-organisms which infest the secretions in these diseases find a favorable nidus in the denuded spot caused by the foreign body or its removal. They multiply, infiltrate the corneal border of the wound and an ulcer results. When the infective process is not a very active one the ulcer does not increase to a large size and does not reach the deeper layers of the cornea, but in certain other cases where the resistance of the tissues is low and the supply of micrococci large and vigorous the most serious and rapid destruction of the cornea may result.

SIMPLE ULCER is usually single, central, small, of a grayish-white appearance and is accompanied by considerable pain and lacrymation. There is a good deal of pericorneal injection and some photophobia. This disease may be distinct from phlyctenular keratitis or it may be one of the single pustules of that disease which has burst and become an ulcer (57). The treatment is practically the same (63), viz.: atropine, rest to the eye, and frequent use of a hot disinfectant lotion (31). General treatment is to be given if needed and any accompanying conjunctival, palpebral or lachrymal disease (49) should not be forgotten.

SPREADING ULCER (*ulcus serpens, infecting ulcer*) is a much more serious disease than the foregoing, although its beginning may be the same. Its chief characteristic is that it tends to spread over the surface of the cornea and to eat into its substance. It presents an excavation filled, or partially filled, with pus, and although its centre is more opaque than the edges, the latter are surrounded by a grayish zone of infiltration. There is always considerable swelling of the conjunctiva and injection of both the deep and the superficial vessels. If allowed to go on the ulcer increases in size, and there may be much (although occasionally there is very little) photophobia, pain, and lacrymation. An abscess is now very likely to form in the deep layers of the cornea (*onyx*), and a stream of lymph, mixed with escaped pus cells, slowly trickles down from it into the anterior chamber, forming a yellowish-white collection in its inferior segment, as indicated by Fig. 10. This

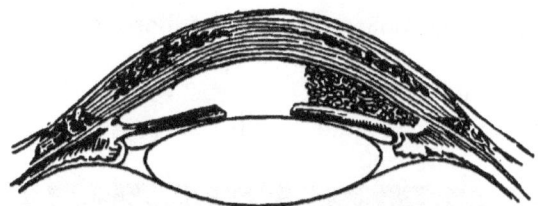

FIG. 10.

condition is termed *hypopyon*. Later on the anterior chamber may gradually fill with pus, but before it is completely full the cornea perforates, the purulent collection escapes, and an attempt at repair follows, as in any other abscess. But the inflammatory action may involve the iris and ciliary body, and eventually destroy the whole eye. When a perforation occurs in the way described the iris almost always becomes entangled in the wound, and if the ulcer heals there remains a thick, opaque scar enclosing a larger or smaller portion of the iris. This is known as *leucoma adherens*. There is then no useful vision, although an artificial pupil (93) may improve matters considerably.

FIG. 11.

Treatment.—As soon as an ulcer is found to be spreading however slowly, the infecting organisms, whose multiplications are the cause of the disease, should be destroyed. There are many methods of accomplishing this end, but the best, quickest, and least painful is the cautery (31), the electro-cautery (Figs. 11 and 12) to be preferred. A good handle and special points are made by the McIntosh Battery Co.

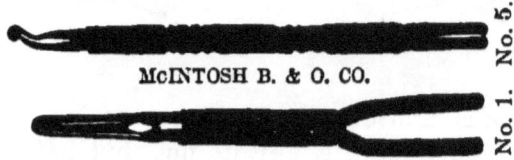

FIG. 12.

The eye is first cocainized and the ulcer well cauterized, no harm being done if the bottom of the ulcer is perforated by the cautery point. Noyes thinks that scraping the ulcer with a spud (Fig. 9) is effectual, and does not leave such a thick scar as the cautery, but other authorities, like Swanzy and Schöler, prefer the latter. After cauterization, impalpable boracic acid powder should be blown into the eye every three or four hours, or hot boric lotion used more frequently. For the boric acid, mercuric perchloride solution (1:5000) may be substituted. It used to be the practice to cut through the whole thickness of the ulcer with a Beer's knife—a proceeding called *Saemisch's section*—for the purpose of evacuating the hypopyon and to lessen the pressure in the anterior chamber. If after a few days following the cauterization the pain is not less, or if the ulcer seems about to perforate, or if the collection of lymph pus in the anterior chamber does not begin to disappear by absorption, a paracentesis should be done.

This is a simple proceeding, and usually carried out by means of a special needle. It is shovel-shaped, and provided with a stop or shoulder to prevent its being pushed too far into the chamber and so injuring the lens. The point of the needle is first directed at an angle of 45° into the cornea, and then pushed slowly through. Once entered as far as the shoulder, the handle should be depressed until the point touches the posterior corneal surface, and then slowly withdrawn. The aqueous gradually flows off. The tapping may be repeated as often as is necessary. It may be made through the centre of the ulcer or at the bottom of the chamber. In either case, more or less hypopyon matter, if present, will escape.

HERPES OF THE CORNEA has already been spoken of (37). The clear bead-like vesicles that first form are rarely seen. They soon break, and their place is taken by irregular spots of disturbed epithelium, easily detected by the reflex test (7).

SENILE ULCER, *concentric* or *ring ulcer*, is most commonly seen in persons whose nutrition is low—old people especially. It travels slowly, is confined to the margin of the cornea, and may heal at one end while progressing at the other.

Treatment.—The most important point is to improve the general condition and to increase the assimilating powers. Paracentesis (71) through the ulcer is indicated, and it should be followed up by the local application of hot sublimate solution (1: 10000). Sometimes, in spite of all treatment, the disease goes on until the eye is lost. When this takes place, a badly nourished organism is to blame. The disease is in reality a senile gangrene of the cornea.

SEQUELÆ OF ULCERS are, so far as the cornea is concerned, *facet, astigmatism* (R. 42), *opacities*, and *staphyloma anterius*. The two former have to do with irregularities of the corneal surface produced by the disease. Instead of a regularly round and smooth surface, some ulcers in healing leave a transparent but faceted spot which greatly interferes with good sight, and may, if central, damage vision by producing irregular astigmatism [C. 21, R. 42]. The healing of a peripheral ulcer may also change the shape of the cornea and bring about astigmatism.

OPACITIES OF THE CORNEA are very common, and may result from any of the diseases which interfere with its nutrition. Granular lids (M. 45), ophthalmia neonatorum (M. 81), gonorrhœal ophthalmia [M. 86], the various kinds of keratitis (56), ulcers, etc., are fruitful sources of them. It goes without saying that central opacities interfere more with sight than peripheral blemishes. Even when they are very faint—so faint as to be scarcely visible to the unassisted eye (9)—vision may be lowered by their pressure from $\frac{20}{20}$ to $\frac{20}{40}$, or one-half. When of this description, faint and hazy, they are termed *nebulæ*. A more opalescent, less translucent scar is called a *macula*, while a dense, white, and quite opaque cicatrix goes by the name of *leucoma*.

In children, especially when the opacity is recent, it is wonderful how much can be done to remove opacities of cornea—even maculæ. The older the patient and the cicatrix the less the probability that the scar will be removed.

Treatment. That the absorption of scar tissue may take place it is advisable that the blood supply to the cornea should be larger than normal. In recent cases of ulcer it is well to prevent the atrophy and disappearance of the blood vessels of repair which run in the corneal tissue to the lesion from the conjunctival margin. Both these objects are best attained by the local use of remedies calculated to slightly irritate the cornea and conjunctiva. Of these by all means the most efficient is massage with the oxide of mercury, or citrine ointment (27). A little should be placed in the conjunctival sac and thoroughly, though gently, rubbed once or twice daily and for five minutes at a time upon the corneal surface through the closed eyelids. This system of *massage* is very useful in many cases of chronic diseases of the conjunctiva and cornea. Another plan (Berry) is to put into the eye once a day a drop or two of equal parts of turpentine and olive oil. If these measures fail to bring vision up to $\frac{20}{200}$ an optical iridectomy (93) is indicated. The iridectomy itself often assists the absorption of the opacity in some mysterious way. It should be made, if possible, in the lower-inner quadrant, since rays of light from both near and distant objects reach the macula more perfectly throughly an artificial pupil made here than in any other part of the iris.

However, the greatest amount of opacity may be in this situation and then the lower-outer quadrant stands next in order of preference. The two upper quadrants are partly covered by the upper lid and are least desirable.

STAPHYLOMA ANTERIUS. The scar tissue resulting from a deep ulcer is not as resistant of intra-ocular pressure as the normal cornea. It sometimes happens that the weak cicatrix gives way under this pressure, stretches and produces an unsightly bulging forward of the cornea. This bulging may become so great that the lids cannot be closed over it.

Treatment. An iredictomy should first of all be performed, in the hope, as sometimes happens, that it will arrest the progress of the deformity. If it does not enucleation (132) or evisceration (the removal of the entire contents of the globe) must be done.

TATTOOING THE CORNEA is an efficient method of covering over unsightly white leucomata or maculæ. The best India ink (in the form of paste) should be used, the cornea must be well cocainized and the coloring matter is driven obliquely into the scar by means of two or three fine needles mounted in a handle or firmly set in a piece of cork (Fig. 13). Two sittings are usually enough to complete the work, which will have to be repeated every six months or every year.

FIG. 13.

SCLERITIS AND EPISCLERITIS. In these diseases (and it is difficult to separate the one from the other) there is scleral injection, pain (severe in some cases, almost absent in others) and swelling of a purplish color at the point affected—usually 3-4 mm. from the sclero-corneal junction.

It is not a common affection and is likely to be mistaken for conjunctivitis or iritis. A little care will detect the circumscribed reddish swelling or swellings which characterize the disease.

After recovery dark pigmented patches often remain to indicate the site of the acute lesion. It is more frequent in women than in men, is often obstinate and chronic and is nearly always caused by the poison of rheumatism.

Treatment. Hot fomentations, atropine, goggles and, when there is much pain and pericorneal injection, leeches (21). The rheumatic taint should be neutralized by potassic iodide, sodic salicylate and other appropriate remedies.

When the acute symptoms have been relieved, great benefit may be derived from massage.

LESSON VI.

DISEASES OF THE IRIS AND ANOMALIES OF THE PUPIL.

Coloboma of the Iris—Differences in Color—Albinism—Nystagmus—Iridodialysis—Various Kinds of Iritis—Iridectomy—Variations in the Size of the Pupil—Myosis and Mydriasis—Various Conditions which Produce Anomalies of the Pupil—Hippus.

It sometimes, though rarely, happens that a section of the iris is congenitally absent. This condition, which resembles that produced by iridectomy (93) goes by the name of COLOBOMA. It occurs in the lower-inner quadrant of the iris and is usually accompanied by other deficiences within the eye and about the head. Vision is not much affected by it.

CONGENITAL DIFFERENCES OF COLOR (*heterophthalmos*) are occasionally to be seen. One iris may be blue, while its fellow is brown, or a portion of the same iris may be one shade or color and the remainder of quite a different hue.

ALBINOS, condition *Albinism*, have little or no pigment in the iris and choroid. These persons have white hair and pink irides. They "screw up" their eyes to exclude the light because the choroidal pigment, which prevents excessive retinal irritation from unabsorbed light waves, is absent. They strive to exclude a portion of the light by reducing the aperture to a mere slit.

NYSTAGMUS is a common accompaniment of albinism. It may be described as a spasmodic jerking of the eye-balls of nervous origin and is often witnessed in affections of the retina and optic nerve.

Treatment. Any defect of refraction should be remedied, and tinted spectacles, having a narrow slit in them (*stenopaic* glasses), may be ordered. The patient should of course be first tested with a stenopaic obturator from the trial case (R. 8) to discover in which meridian he sees best.

IRIDODIALYSIS, or the separation of the iris from the ciliary body, is always the result of blows upon the eye. More or less bleeding into the anterior chamber (*hyphœma*) accompanies this lesion. The use of atropine is indicated in the hope that the edges of the wound may unite.

IRITIS. Inflammation of the iris. The following account* will serve to indicate the salient features of the several varieties of iritis—one of the most important ocular affections the practitioner has to deal with.

The history which G. W., æt. 22, gives of himself, as he comes in with a bandage over his left eye, is the following: Six years ago he had an attack of "inflammation" in both eyes, and was treated for it with "eye water." Since then he has had three similar attacks, the last one (for which he now presents

---

*Abstracted from a clinical lecture by the author on "Ordinary Forms of Iritis," from the *North American Practitioner*, July, 1890.

himself for treatment) affecting him as did the others. His eyesight each time after recovery from two of the attacks has been noticeably weaker than before. On inquiry the patient denies that he has or ever had any venereal disease (and there are no signs of it about his person), but he has a distinct history of rheumatism. The eye affection first followed upon an attack of acute rheumatic arthritis, with which he was laid up for two months, and he has since then had several attacks of the same disease. Regarding the present attack, he says it resembles the others, only that the general rheumatic symptoms are very slight. It has already lasted three days. He complains of considerable pain in the supra-orbital region, and he says the whole side of his head aches. His left eye is decidedly "red;" it "waters" a good deal; there is considerable photophobia, and the pupil is smaller than on the right side. His eyelids do not stick together, and there is no discharge of pus or muco-pus from the eye. He also complains of pain on touching the eyeball. He makes no complaint about the right eye. We shall put a few drops of a 4-grain solution of sulphate of atropia into both eyes, have him wait half an hour or so, and notice the result.

The disease from which the patient suffers is iritis, probably of rheumatic origin, and the importance of a correct and *early* diagnosis in all of this class of cases is so great that it would be well to ask ourselves: (1) What are the most reliable and most useful signs and symptoms of inflammation of the iris? (2) What diseases is one most likely to mistake it for, and how can it be distinguished from these affections? (3) What will probably occur if an early diagnosis is not made, and efficient treatment resorted to? (4) What treatment accomplishes most good in the several varieties of this inflammation?

As a necessary preliminary to these questions, one might further inquire, What is the essential nature of iritis—its pathology? Simply that of inflammation in a highly vascular structure. The iris is a mixture—speaking roughly—of a small percentage each of unstriped radiate and circular muscular fibers, nerve fibers and endings, ganglia, brown pigment, and lymphatics, interspersed with a larger proportion of blood-vessels and lymphatics—a fine field for inflammatory processes to run riot in. Its anterior surface is uneven—dotted over, here and there, with small hills and valleys, all of which are clothed with the color-giving pigment. Remember, too—for it has its place in diagnosis—that everybody's irides, it matters not what color they may appear to the observer—blue, brown, gray or black—contain the same kind and almost the same amount of this brownish coloring matter. It is its variety of arrangement on the iris surface that gives rise to the different color impression.

It is possible, I think, to trace the "cardinal signs" of inflammation in such a case as we have examined. Inflammatory *redness* is not developed in the iritic tissue itself, but is found at the corneal margin. Enlarged vessels, which are the cause of the sign of redness in inflammation of tissues, are, without doubt, present *in* the substance of the iris, but, for obvious reasons, they are not manifest.

What one does see at the commencement of an attack—and it is well seen in our patient—is a faint zone of redness, about 5 mm. in width, encircling the cornea. If we look at the eye with a magnifying glass, this pinkish circlet is seen to be due to a number of small and almost straight vesssls, which do not lie in the conjunctiva, but are under it. This is proved (6) by gently pushing the overlying mucous membrane to one side; these fine vessels do not move with it. Not much reliance, however, can be placed upon redness as a diagnostic sign when the iritis is severe, because all the vessels of the conjunctiva, iris and sclera become involved in the inflammatory process. If one studies the vascular supply of the eyeball, it will be seen that the iritic vessels anastomose with those of the conjunctiva on the outside of the globe, and form part of the uveal plexus within. *Change in color* is somewhat allied to the sign of redness, and in the case under observation is so marked that even the most casual observer would be sure to remark it.

The patient has a mud-colored left eye and a blue right one. When inflamed, the iris loses its delicate velvety gloss and tracery; blue eyes look clay-colored, brown eyes assume a greenish hue, and so on. When both eyes are affected this is not so marked, as there is no contrast between the two sides.

Swelling is a sign not easily made out in iritis, but its presence brings about a very noticeable feature of the disease—a contracted pupil. The infiltration of the loose tissues of the iris by inflammatory products causes the pupillary margins to approach, and the pupil becomes smaller.

Pain is almost always present. It varies greatly in intensity, aud is one of the characteristic elements of the disease. It usually increases in violence towards evening, eases off in the night time, and may get worse again in the early morning. It affects preferably the temple, vertex, and the globe itself—in reverse order.

The supra-orbital twig of the trigeminus is the efferent nerve chiefly affected, and even in such a mild case as we now have on hand, pressure upon it at the supra-orbital notch and in its course over the forehead reveals several sensitive points.

Likewise, the iris suffers when inflamed, from *impairment of function;* the pupil does not dilate when the eyes are shaded from the light, nor contract to any extent when suddenly exposed to it.

Iritis, according to the severity of the attack, may last three or four days, or it may continue for months. The globe is nearly always tender on pressure while the disease lasts. This sign alone is often sufficient to distinguish it from some diseases that resemble it. Of these, by far the most important are conjunctivitis, neuralgiæ of head and face, and the various forms of inflammation of the cornea.

One of the commonest and most unsatisfactory experiences of the ophthalmologist is to be called upon to deal with an old iritis, which has been treated for conjunctivitis. Such mistakes ought to be less frequently made than they are, because, beyond the scleral redness caused by injection of the vessels overlying that structure, *iritis and conjunctivitis have nothing in common.* I would advise the non-specialist to disregard entirely the matter of vascular injection in the diagnosis of external eye diseases. It is entirely untrustworthy, and, as compared with other signs and symptoms, of little value. For the sake of comparison and contrast, let us picture side by side the salient features of these affections:

CONJUNCTIVITIS.		IRITIS.
Muco-purulent or purulent, causing morning adhesion of lids.	*Discharge.*	Watery. Does not cause lids to adhere.
Comparatively little; if present, mostly confined to globe.	*Pain.*	Often severe and neuralgic in character. Worse at night and in the early morning.
Not affected.	*Vision.*	Affected often and early.
Not much; often none.	*Photophobia.*	Nearly always present.
Dilates when eyes are shaded.	*Pupil.*	Dilates sluggishly or not at all.
Both eyes.	*Disease affects.*	May affect one only, or one at a time.
No change in color or appearance.	*Iris.*	Discoloration, with loss of velvety gloss.

The fact does not appear to be generally known, but it is a fact nevertheless, that cases of iritis are sometimes treated as hemicrania, supra-orbital neuralgia, facial neuralgia, malarial headache, etc., and neither patient nor physician is meantime aware that a serious intra-ocular inflammation (of which the pain happens to be the most urgent symptom) is "blazing away" unchecked. The "red eyes" are in such cases attributed to "congestion of the head," "hyperæmia of the conjunctiva," or to some such cause. Of course, inspection of the iris would correct this error of diagnosis.

There is only one way to separate corneal troubles from iritis when (as generally happens in keratitis) intolerance of light is a prominent symptom, and that is to put into the affected eye two drops of a 10-grain solution of cocaine every couple of minutes for a quarter of an hour. This will quiet the eye, and enable one to obtain a good look at the cornea and iris. One thing is, however, worth emphasizing in this connection: If the patient is young—under twelve years of age—the photophobia is almost certain to be due to one of the forms of inflamed or ulcerated cornea, and not to iritis.

Now let us return to our patient. I find that the atropine has dilated one of the pupils quite wide, but the other resists the mydriatic. I have made a sketch of both pupil and irides, so that the different effects of the drug upon them may be seen.

Why have not these pupils fully dilated, and what do their irregular margins mean? If one looks closely at the right pupil—illuminating it with a not too strong light—he will observe that a portion of the iris is attached behind to the lens. It looks as if a tack had been driven through the edge of the iris into the lens, and so prevented it from being drawn back with the remainder of the organ.

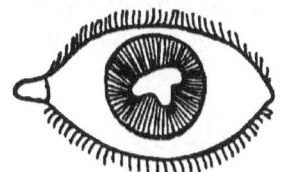

FIG. 14.

Such a state of things is more marked in tne left eye. Here, indeed, it looks as if the receding iris were composed of two layers, an anterior and a posterior, capable of some degree of motion upon one another, and that the posterior were "tacked" at several points around its margin to the anterior capsule of the lens. When the whole curtain of the iris was drawn back by the atropine, the anterior layer seems to have yielded in its entirety; so has the posterior, except at those points where it was not adherent to the lens surface. And this is about what has happened. The swollen iris, during some previous attack, has (as it always does in iritis) come in contact with the lens, and inflammatory adhesions have formed between the latter and the posterior surface of the iris. These adhesions—posterior synechiæ, they are termed—at first soft and easily broken up, have become organized and fibrous—just as occurs elsewhere—and a damaged eye is the consequence.

As a rule, the more frequent the attacks, the greater likelihood of connective tissue bands forming in the way described; but, on the other hand, a single attack of iritis, unless properly treated, may leave an eye in the condition of our patient's left. Some small, dark spots are also to be seen in the background of his widely dilated right pupil. These are small dots of pigment, which have been torn from the posterior iritic layer while the inflammatory exudation was recent. It often means that atropine has been used, and that it is the resulting mydriasis which haş produced this effect; and although pigmentary deposits on the surface of the lens sometimes reduce the acuteness of vision when they are close to the centre of the pupil, they are less objectionable than an iris glued to the lens capsule. As a result of these adhesions, the patient may suffer not only from interference with vision consequent upon a partially obscured pupillary aperture, but the adhesions may, if extensive enough, give rise to more serious troubles. Indeed, they may lead to actual destruction of sight, from certain secondary changes in the eye (cyclitis, glaucoma (113), etc.), which we cannot discuss just now. It is very probable, also, that the continual tugging at the little points of adhesion (as the iris attempts to expand and contract) may dispose the damaged iris to recurrent attacks of the disease.

The iris may become inflamed as the result of injury, but by far the commonest causes of this condition are syphilis, gonorrhœa, and the poison of rheumatism. It is, of course, very important to get at the cause of the iritis in a particular instance, as upon that depends in a great measure the successful treatment of the case. In syphilitic iritis the inflammation is usually more plastic than in the rheumatic or traumatic forms of the disease, and it may show itself as a secondary, a tertiary, or even as a congenital manifestation. It probably affects both eyes more frequently than the rheumatic form. The latter is, however, the most obstinate, and the most liable to recur of these forms. In either instance proper constitutional treatment should be employed.

The pain, if not severe, is often relieved by hot applications. Some patients like dry heat; others prefer wet applications.

Often two full doses of antipyrin, given at intervals of two hours before the usual exacerbation, act magically, and secure the desired sleep. Cupping the temples, or the application of leeches, is good practice, and is often resorted to.

But of all the remedies in the pharmacopœia, sulphate of atropia (or, if it causes too much local irritation, (32) some salt of duboisia in the same dose) is by all odds the best. It matters not at what stage of the disease one sees the patient, one should make an effort (by instilling into the affected eye, every hour or two, a few drops of a 4-grain solution) to dilate the pupil. When that is accomplished, every four or five hours will be sufficient. My practice is to begin by putting in, *myself*, once a day, until as complete a dilatation as possible is accomplished, as much as can be retained in the conjunctival sac of an 8-grain to the ounce mixture of atropine and vaseline. The effects of this mixture are more lasting than the solution, and it does not, I think, run off or get so readily into the nasal passages and produce constitutional effects. The patient meanwhile uses a weaker mixture or the solution aforementioned—preferably the solution—unless you can feel certain that some competent person will apply the ointment for him.

The use of atropine accomplishes several things. It dilates the pupil, and so prevents the dreaded adhesions between lens and iris. It relieves the neuralgic pains, promotes absorption of the inflammatory exudations, reduces the capillary congestion, and probably cuts short the disease. Even when synechiæ have formed, it may tear them asunder, if the case has not been seen too late.

So potent is this drug that, speaking generally, one may affirm that when the pupil has been dilated by a vigorous use of atropine, and is kept dilated by smaller doses of the same remedy, the disease is under control, and recovery soon follows. We shall also order a shade for this patient, which he must wear over his eye while in the house. It is best, for many reasons, that he should abstain from work, and not use even his, as yet, unaffected eye. When he comes again to see us, it must be on a fairly mild day, and he ought to wear a light porous bandage (without a pad) over the left eye.

His rheumatism should not be lost sight of. I think the present attack of iritis should be considered as a sub-acute manifestation of the general dyscrasia. We must treat it as such by appropriate remedies, well known to the profession.

When syphilis has been the cause of the disease, the character and dose of the constitutional remedies will largely depend upon the relation which the iritis bears to the initial lesion, whether the former be a secondary, a tertiary, or a congenital manifestation. Overdosing with powerful mercurial remedies is always to be deprecated. Inunction with ungt. hydrarg. once daily—with intermissions of three days after each week's use of the ointment—combined or not with potassic iodide internally, will be found to act well. In gonorrhœal iritis the eye symptoms experience considerable relief when the urethral discharge is stopped.

IRIDECTOMY.—When the iritis is of the recurrent form a broad iridectomy often prevents a return of the disease or lengthens the interval between the attacks. The operation itself is performed in the following manner: The conjunctival sac having first been irrigated (31) with an antiseptic solution, and all the other antiseptic precautions carried out (31), a

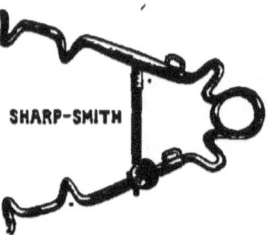

Fig. 15.

4 per cent. solution of cocaine is dropped into the eye. The wire speculum (Fig. 15) is now introduced, and the eye being steadied by fixation forceps (Fig.

Fig. 16.

16), an incision is made at the sclero-corneal junction with a narrow Graefe knife (Fig. 17). This cut should

Fig. 17.

be made in the same way as for cataract (105), but should not be as wide as the latter. Or a keratome (Fig. 18) specially devised for this purpose may be employed. If the latter be chosen, it is first entered at right angles to the globe, and when the cornea is pierced the handle is depressed and the point pushed

FIG. 18.

across the anterior chamber as deeply as necessary. The aqueous begins to escape as soon as the withdrawal commences, so that the sharp point of the in-

FIG. 19.

strument should now be made to hug the posterior surface of the cornea, else the lens may be wounded. The width and situation of the corneal opening should

FIG. 20.

correspond with the kind and size of the iris coloboma desired, a narrow one for small, optical iridectomies, and a wide and very peripheral one for recurrent iritis (91) or glaucoma (113). (See Fig. 19). The fixation forceps is now given to the assistant, and with iris scissors (Fig. 20) in the right hand, iris forceps

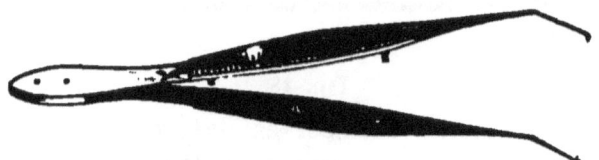

FIG. 21.

(Fig. 21) are, with the left, entered closed and carried along the posterior corneal surface until the pupillary margin is reached. Here they should be opened, the

FIG. 22.

iris grasped, pulled well out of the wound, and snipped off with one cut of the scissors (see Fig. 22) close to the cornea. This part of the operation is the only one attended by pain, and the patient should be warned of it and assured that it will not be severe and will last for an instant only. Care should now be taken that the edges of the cut iris are replaced either with the repositor or, what is still safer, by gently stroking the cornea over with the rubber spoon (Fig. 23).

FIG. 23.

THE SIZE OF THE PUPIL varies greatly in health and disease. For example, there is a considerable degree of physiological variation due to the different amounts of light which at different times reach the retina. Also, as is well known, the pupil contracts when the eye accommodates (C. 24) for near vision, while it expands on again looking in the distance.

MYOSIS, or tonic contraction of the pupil, is a very important symptom of disease of the spinal cord,* and as such is commonly called *spinal myosis* (125). It is also produced by drugs, such as eserine, opium, or by any local irritant to or disease of the iris.

MYDRIASIS, or dilatation of the pupil, is most commonly due to the action of atropine, belladonna, or some other mydriatic (22), but may result from paralysis of the contractor fibres of the iris, due to disease. Thus we find it in paresis of the third nerve (137), in glaucoma (113), in optic atrophy, after diphtheria (125), and in several other nervous disorders.

---

* Hypermetropic individuals (when the refractive error is considerable) and old people have small pupils.

In diphtheria this paretic symptom, like that affecting the pharyngeal muscles, tends to get better without treatment. The pupils of the myope and of hysterical women are larger than normal.

HIPPUS.—This term describes an alternate contraction and expansion of the pupil, such as one occasionally sees in cases of nystagmus (80).

LESSON VII.

CATARACT AND OTHER AFFECTIONS OF THE CRYSTALLINE LENS.

Dislocation of the Lens—Irido-donesis—Aphakia or Absence of the Lens—Cataract—Nuclear and Cortical Cataract—Senile Cataract—The Operation for Removal—When to Operate—After-Treatment—Complications—Soft Cataract—The Zonular or Lamellar Cataract of Children—Discission or Needling.

DISLOCATION OF THE LENS. The crystalline is beautifully swung by means of its suspensory ligament, or zonula of Zinn, from the ciliary body and is so arranged as to withstand the influence of ordinary jars or injuries to the ocular region. But severe blows, delivered directly upon the globe may, particularly if the ligament be weakened by disease or if the vitreous be fluid (117), cause rupture of some of the suspensory fibres and the lens may thus become *dislocated*. When the zonula is torn to a slight degree only, the dislocation is usually correspondingly slight; but when the violence is considerable and directed towards the ciliary region, the lens may be torn entirely away from the ligament and driven into the vitreous, into the anterior chamber or even through the burst sclerotic underneath the conjunctiva. In every case there will be observed more or less trembling of the iris, or *iridodonesis*, when the patient is told to look in various directions, and this is diagnostic of the injury. Shortly after dislocation, through interference with its nutrition, the lens grows hazy and may finally become quite opaque. If the pupil be dilated with homatropine, or better still with cocaine and atropine, the lens may be made out by means of reflected light (20) and its new position definitely determined.

The power of accomodation (C. 24) is lost when the crystalline is dislocated, since the accident necessarily interferes directly with the mechanism of the act. For the reason which is given in the description of aphakia (98) the eye also becomes very hyperopic when the lens is no longer in the axis of vision.

Treatment. The best treatment of a *slightly* dislocated lens is to leave it alone, unless it becomes so opaque as to decidedly interfere with vision. Where the dislocation is very marked, or where the lens is loose and "bobs" up and down in a fluid vitreous, or where it is dislocated into the anterior chamber, it should, as a rule, be removed. The successful management of these cases requires some ophthalmological experience, and the operation of removal often calls for the exercise of considerable skill and dexterity.

APHAKIA. Whenever the lens is absent, whether as the result of operation, absorption, or dislocation, the condition is styled aphakia. The refractive power of the crystalline being equal to about 10 diopters (R. 8), the eye is made hyperopic to that extent, and after cataract operations, as is well known, this has to be taken into consideration and suitable glasses ordered.

Moreover, as aphakia necessarily involves a loss of accomodative power (R. 13) a glass for reading and other near work is required in addition to the distance lens.

CATARACT, or opacity of the lens, is the result of structural changes in the fibres of which it is composed. Sometimes these alterations of structure are irregularly distributed throughout the body of the lens, as in most cases of senile cataract; sometimes they are mainly central (*nuclear* cataract) sometimes they are confined to the periphery or cortex (*cortical* cataract).

There is a great variety of cataract, but the most practical division of them is into "soft" and "hard" cataract. Speaking generally, the soft variety occurs in persons below thirty or thirty-five years of age, while the hard variety is found only in persons above that age.

SENILE CATARACT. This is the commonest as well as the most important form of the disease. The normal lens gradually undergoes changes as we advance in years.

In old age the lenticular nucleus becomes firmer, and with the rest of the lens acquires a yellowish tint and transmits less light than formerly. When the pupil is widely dilated the grayish, translucent outline of the whole crystalline is distinctly visible both by the oblique (9) illumination as well as by reflected light (19). While these changes attendant upon old age may be regarded as more or less physiological, they are closely allied to true cataract. In the latter instance, however, delicate lines, or well defined streaks of opacity, best seen by reflected light with a dilated pupil, add themselves to the gray tinge of the lens. These commonly begin at the periphery, or *equator*, and gradually invade both cortex and nucleus until the whole crystalline is involved. The pictures made by progressive senile cataract, from the time when the first faint dark lines appear until the cataract is complete, are often very pretty and remind one of the geometrical shapes one sees when snowflakes are examined by a lens.

The *time* occupied by the process varies greatly but may extend over many years.

Symptoms. Sight will not be much affected until the nucleus is involved. The patient then complains that images are distorted or multiplied (*polyopia*), or that there is a cloud or floating bodies before the eyes. This fogginess increases very slowly until finally the visual acuity (12) is reduced to the counting of figures.

When cataract is *ripe* or *mature* the whole lens is opaque and it can be safely removed by operation. Such a cataract should have a regular mother-of-pearl appearance by the oblique illumination (9), and while this examination is being made (the pupil *un-dilated*) the iris should not cast a shadow on the lens surface. There should also be no glittering sectors (Förster) or facets brought to view as the patient is told to look in different directions during the examintion. When an operation is undertaken on an immature cataract soft matter is almost certain to remain behind. These small masses, when left behind, are not only liable to set up iritis, but, transparent at first, finally become opaque and lower the visual acuity. Finally, there should be no "red reflex" (20) to be seen.

When both lenses are cataractous the patient is doomed to go about in a condition of practical blindness for months unless the cataracts are *artificially ripened*. The only effective method of bringing this about is the plan of Förster. A preliminary iridectomy (92) is made and the capsule is gently and carefully rubbed over the cornea by a rubber spoon (95). Or if it be decided not to do an iridectomy (8) the same massage may be accomplished by first tapping the anterior chamber (73). In a few weeks the lens will be found to be mature.

REMOVAL OF SENILE CATARACT. Before attempting the operation of extraction—the only efficient treatment of this form of cataract—certain important precautions must not be forgotten:

1. The cataract should be mature.
2. The patient's health should be fairly good,

else the healing process may be interfered with. As hard cataract occurs in old people, who are not as a rule robust, we may have to deal with some cases who are not encouraging subjects for any kind of surgical treatment. It is wise to make the best of such patients and to improve their condition as far as it is capable of improvement. *Cough, constipation,* and *insomnia* should be relieved if they be present.

3. The external eye should be examined. Mucocele (50), blepharitis (34), conjunctival (M. 15) and corneal diseases should be treated and, if possible, cured before operation. If this is neglected, the abundant germs which these diseases supply are very likely to infect the corneal wound and lead to dreaded complications.

4. Corneal opacities should be searched for, and, if found, the patient should be warned that they form a bar to realization of perfect vision.

5. It is very important, in view of their ultimate effect upon the visual acuity, that the presence or absence of deep-seated disease of the eye should be demonstrated. It would be very disappointing, after an operation entirely successful from a surgical standpoint, to find that the patient had had all the while disease of the optic nerve, for example, and was unable to see any better after the extraction than before it. Such a case came under my observation, not long ago, where the examinations about to be described were omitted. In a case of *uncomplicated*

cataract, then, the patient should have prompt perception of light—should, in other words, be able to state at once when the hand is passed over his eye between it and a good window light. In a dark room he should be able to see a candle flame 20 feet away, or a faint light reflected from a mirror (20) at a metre's distance, and to point out its locality when moved about in different directions. This "projection of light" test should never be neglected. It is really a test of the perceptive ability of the different sections of the retina, and fails when any considerable part of the latter is diseased.

There are almost as many forms of cataract operations as there are operators, but the usual method of extracting senile cataract is the so-called *modified peripheral linear operation* of Von Graefe. The instruments needed are a pair of fixation forceps (Fig. 16), a wire lid speculum (Fig. 15), a narrow Graefe cataract knife (Fig. 17),—whose well sharpened point pierces readily, and by its own weight, the testing drum (Fig. 24), a pair of iris forceps (Fig. 21), a metal or shell iris repositor, a pair of iris scissors, and the cystitome (Fig. 25.)

Examine them with a lens and be sure that they are perfectly bright, sharp, and clean. One drop of a freshly prepared 4 per cent. solution of cocaine is dropped into the eye every minute for five minutes. The speculum is now introduced by sliding the upper branch under the upper lid, and then the lower end under the lower lid. It should not be opened too widely lest it cause pain.

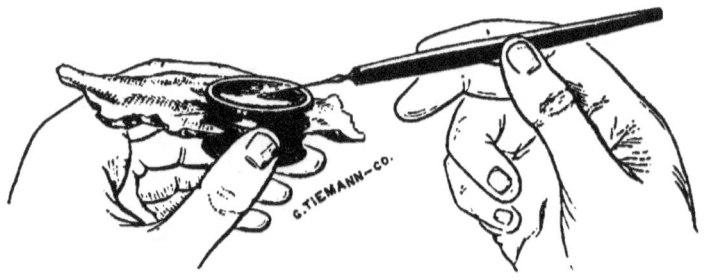

Fig. 24.

The patient should now be warned not, under any circumstances, to "squeeze the lids together" during the operation. It would also be well to exercise him beforehand in looking down, up, in and out. Whether it be from "operation-terror" or what not, some patients cannot be induced to perform these simple acts at those critical moments when they are urgently called for. A little preliminary drill will be found useful in such cases.

Fig. 25.

The first step is the corneal incision. This should be made at the limbus in the upper corneal semicircle, and will involve about two-fifths (better more than less) of the whole circumference, as in Fig. 26. The patient looking up, the conjunctiva and subconjunctival tissue are grasped as represented. He now looks down, and the point of the knife, edge up, is entered, and is directed downwards. The handle is depressed and a counter puncture is made, and, by a to and fro motion, the blade cuts its way out, as along the dotted line of the diagram.

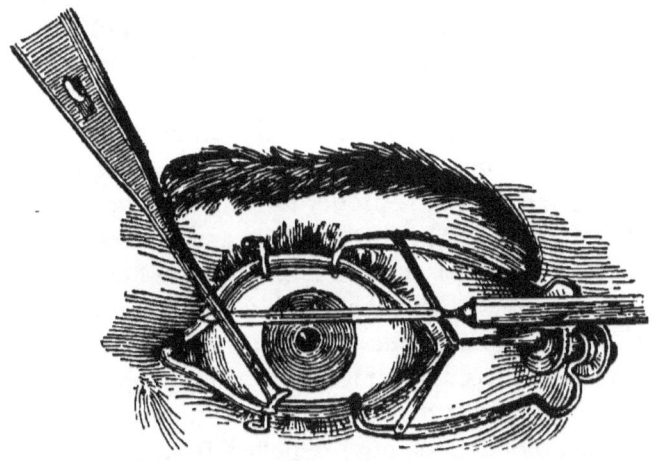

FIG. 26.

The second step—the iridectomy—has been already described (92).

The third step is the opening of the capsule to allow of the escape of the lens. As soon as all bleeding stops the patient looks down, the surgeon again fixes the globe, the cystitome (Fig. 25) is introduced, and crucial scratches are gently and carefully made on the cataract surface.

The fourth step. After the cystotomy the lens will probably present in the wound, and its complete delivery may be accomplished (the patient looking down) by gently pressing, midway between the centre and the lower edge of the cornea, with the rubber spoon. The pressure should be directly backwards, and no attempt ought to be made to squeeze out the lens by even the appearance of force. Loss of vitreous is apt to occur unless this precaution is observed. The opaque lens may, however, be "followed up" by the spoon as it emerges from between the lips of the wound. Here the assistant will loosen the fixation forceps, and the speculum had better be removed.

Fifth step. It is proper to coax out, by stroking the cornea from below upward, any masses of soft matter or pieces of capsule that may have been left behind. The pupil, which was before white, is now black, and these remains can usually be seen. When the edges of the iris coloboma are in their proper places, and everything (iridic, lenticular, and capsular remains, vitreous, blood-clots, etc.) removed from between the edges of the wound—some surgeons employ boric acid irrigation for the purpose—the lids are gently closed.

DRESSINGS innumerable have been recommended after cataract operations. While it is well not to follow empirically any one plan, I would advise the following: A small piece of old, aseptic linen, cut so as to fit the oculo-nasal angle, is thinly spread with this ointment:

Finely powdered boric acid............	ʒ j.
Atropine sulph................. ⁙...	gr. ij.
Cold cream.....................	ʒ j

It is then evenly applied to the closed lids. Next, a thin layer of borated cotton, and over all a flannel bandage (29). If the patient will keep moderately quiet, there is no necessity for confining him to a dark room, or even putting him to bed. A shade over the eyes, and a comfortable high-backed arm-chair, are much to be preferred to the dark room, confinement to bed, and absolute quiet of the old *régime*.

The first night a dose of sulphonal, 2 grammes (30 grains) two hours before retiring, with ¼ grain of morphia just before the usual hour of sleep, may be administered.

The *after-treatment* is important. During the twenty-four hours succeeding cataract extraction, most patients complain of smarting or occasional twitches of pain. These have no serious significance, and are usually relieved by the morphia given at night. If the eye feels perfectly comfortable the bandage may be left until the third or fourth day. As a rule, however, the patient will be more at ease if his closed lids are bathed cautiously, daily, with warm boric acid lotion, and fresh dressings applied. He is not in that case tempted to rub his itching lids or brow, to the detriment of the healing process, as is sometimes the case when the eye remains untouched for several days. If all goes well, glasses for reading as well as for distant vision can be ordered in about six weeks.

COMPLICATIONS.—Continued dull pain after the first twenty-four hours, or excessive discharge, usually means mischief and should lead to a removal of the bandage and a critical examination of the eye. These signs, when they occur during the first three or four days after the operation, point to the invasion of the corneal cut by micro-organisms. Later on—five to ten days—they mean iritis. In the first instance, purulent infiltration of the edges of the wound is indicated by a grayish yellow appearance along the line of the incision. As soon as it is discovered, the eye should be well irrigated with hot boric acid solution, and the cautery thoroughly applied to the line of infection. The hot irrigations ought to be renewed every three or four hours until the disease is under control. The secondary iritis has the signs and symptoms of the primary form, and should be treated like it (90).

EXTRACTION WITHOUT IRIDECTOMY is an old friend with a new face. Practiced more than a quarter of a century ago, it fell into desuetude, but has lately been revived and has many renowned advocates. It cannot be denied that other things being equal the preservation of a round, central, and more or less contractile pupil is, in cataract extraction, a consummation devoutly to be wished. On the whole this operation is more difficult to perform than the foregoing. Whether this drawback is offset by superior advantages in the way of better vision, a natural appearance of the pupil, absence from complications, etc., remains just now a debated question. Probably, as Noyes suggests, it is better not to do exclusively either operation, and there certainly appears no reason why we may not come to employing the "simple" method for ordinary cases and adding the iridectomy when good and sufficient reasons call for it.

SOFT CATARACT occurs in children and young adults (99). The most important variety is the *lamellar* or *zonular*. This is the ordinary cataract of infancy and childhood and is either congenital or forms soon after birth. It can easily be made out through the dilated pupil, both by oblique illumination (9) and reflected light (20). The opacity effects the lens in layers, does not extend to the periphery, and the visual acuity is sometimes as high as $\frac{20}{30}$ (12).

Lamellar cataract is not progressive like the senile variety.

A history of infantile fits is almost always given by these patients; they nearly all have "strumous" teeth, that is, the enamel of the incisors and canines is likely to be absent from the crowns and upper halves, and (in England especially) they almost all are certain to have been treated during the teething period with mercurials. What the relation is, if there be any, that exists between the cataract, the convulsions, the rachitic teeth and the mercurial treatment it is difficult to say.

In addition to the different forms of congenital and infantile cataract *traumatic* cataract is of the soft variety. The injury brings about rupture of the anterior capsule; contact of the aqueous humor with the normal lens causes its fibres to swell, become opaque, project through the rent in the capsule and sometimes, as in the operation for discission (111), to become partially or totally absorbed.

Diabetic cataract is also of the soft variety.

Treatment. If, as in some cases of lamellar cataract, the patient's vision be sensibly improved by dilatation of the pupil, an optical iridectomy (93) is indicated. If, however, with correction of optical errors, this does not furnish a useful degree of sight, or if the cataract be *total*, needling or DISCISSION is called for. This operation is employed when absorption of the whole lens is sought to be secured.

The pupil having been previously dilated with weak atropine solution, cocaine is instilled (or, in the case of quite young children, chloroform given) and a "stop" cataract needle (Fig. 27) is passed through the

FIG. 27.

cornea 2 mm. from the outer scleral junction. The point is now directed forwards to the centre of the lens and a single vertical cut made in the capsule. The needle is at once withdrawn, little or no aqueous escapes and there is very little inflammatory reaction. In a day or two a portion of the lens will project through the opening thus made. This, through the solvent action of the aqueous, will be absorbed; another piece will protrude, go through the same process, and so on, until in from three weeks to three months the whole lens will have disappeared. Weak atropine drops should be used throughout and a bandage is advisable. Some surgeons prefer, as soon as the lens mass comes forward into the anterior chamber, to remove it by means of a specially constructed suction syringe, but the discission operation alone is usually all that is required.

Needling in Secondary Cataract.—The posterior capsule is, of course, not removed in either of the cataract operations just described, and if at all opaque, its presence may prevent the patient from obtaining good vision. Fine fibrous threads and iritic remains are also occasionally left after the primary operation. A central rent should be made in the opaque curtain, and this is best done by introducing two discission needles—one in each hand—at the outer and inner corneal borders respectively, and having them meet in the centre of the capsule; the handles are simultaneously elevated so as to cut the desired opening. Even when the capsular cataract can only be made out with the lens (9), a considerable increase in the visual acuity is attained by this operation of needling, and it is commonly resorted to. No reaction should follow.

LESSON VIII.

GLAUCOMA.

The Necessity of an Early Diagnosis—Varieties—Causes—Intraocular Changes in the Disease—Signs and Symptoms—Treatment—Iridectomy and Sclerotomy—Eserine.

In the chapter on iritis some stress was laid upon the fact that to mistake that disease for some other affection, conjunctivitis for example (M. 15), as was not uncommonly done, was to fall into an error fraught with disaster to patient and doctor. A similar statement, even more strongly accentuated, might be made about glaucoma. In its acute form it proceeds to destruction of vision in a very short time, and even the chronic types are distinctly progressive in character. It is usually a disease of the intraocular lymphatic system. The interior lymph stream arises, broadly speaking, from the blood-vessels of the ciliary body and iris, flows from the posterior chamber through the pupillary opening, and empties into the canal of Schlemm (8) at the angle formed by the iris and sclera. It is easy to understand how anything, such as pressure upon this canal by the enlarged lens of old age, extensive posterior synechiæ (87), increased secretion of aqueous, dislocation of the lens (97), etc., which interferes with the outlet, or abnormally increases the inflow, of this lymphatic fluid, may bring about a high ocular tension and produce glaucoma.*

The *local effects* of the disease are those of intraocular pressure. The eyeball grows harder, and the tension (y) may so increase that it feels like a small apple under the fingers, barely capable of indentation. The optic nerve, where it joins the globe, is found to be pressed out or excavated, and if the pressure is long continued it atrophies. The lens is pushed forward so as to diminish the depth of the anterior chamber, and the cornea may be œdematous and hazy.

Glaucoma occurs most frequently in persons over 45 years of age, and is generally found in hypermetropes, the small eyeball of the latter being held by some to predispose to the disease.

Diagnosis.—The following symptoms and signs should ever be borne in mind in making an examination of the eyes of persons over 40 years of age, for it must be remembered that the usual (*acute simple*) form of this affection is readily diagnosed if a little care be exercised.

In the acute "congestive" form the eye is red—as in iritis—and nearly all the other symptoms will be well marked. In the chronic forms, which may last for years, there are intervals of remission, during which, beyond a slight lowering of vision, nothing

---

* From γλαυκος, green—referring to the greenish appearance of the pupil.

abnormal can be remarked. Repeated attacks, however, combine to destroy sight, and finally, if the patient live long enough and unless relief is meantime given, the glaucoma becomes "absolute" and complete blindness, through atrophy of the optic nerve, is the result.

1. *The tension* (9) *is increased.*
2. *The pupil is oval and dilated*, and has a greenish tinge.
3. *Vision is impaired*, recently, progressively and rapidly in the acute forms. There will be a history of *periods of improved sight* in the chronic cases.
4. *There is always pain* (in acute cases of a neuralgic character), usually referred to the branches of the fifth nerve. In the chronic cases the pain is dull and may be confined to the globe. These pains often get worse at night.
5. *Rainbows, fogs and haloes about gas, lamp ana candle lights* are seen by the glaucomatous subject. This symptom is produced by the rays of light coming through the œdematous cornea, and is practically the same appearance that one sees on looking at a light through a glass that has been breathed upon.
6. *The cornea is often hazy.* When decidedly so, it will usually be found to be anæsthetic, as proved by touching it with a camel's hair pencil.
7. *The anterior chamber is shallower than normal* (7).
8. *Congestion of the episcleral veins* is rarely absent,

even in chronic glaucoma, and is probably a passive condition, the result of impeded circulation within the eye.

Treatment.—The acute congestive form calls urgently for treatment which will, if given early, be wonderfully efficacious. The pains will be relieved, the vision will improve, and after the glaucomatous "storm" is over the eye may become almost, if not quite, natural again. This treatment is a broad and quite peripheral iridectomy (93). In acute cases—when the eye is painful and sensitive—cocaine is not absorbed and is not an efficient anæsthetic; some surgeons accordingly give chloroform or ether.

Some operators prefer *sclerotomy.* The pupil is first thoroughly contracted with eserine, and an incision like that for cataract extraction (105) is made, but well within the sclera. It is not completed, but a bridge of scleral tissue about 2 mm. in width is allowed to remain. This prevents prolapse of the iris. In sclerotomy a certain amount of drainage is assumed to take place through the scleral scar, and thus to relieve the intra-ocular pressure. The peripheral iridectomy relieves the choked canal of Schlemm.

Two things are especially to be remembered in connection with these operations: First, that operating on one eye is apt to *precipitate* (not produce) an attack in the other eye; and, second, that the relief given to the intra-ocular pressure may be the cause of bleeding (into the interior chamber and vitreous) from weak and (now) unsupported vessels. Vitreous hæmorrhage is to be feared, as when it is extensive it may lead to final destruction of the eye.

Bleeding into the anterior chamber is not of serious importance. Chronic and subacute glaucoma is often treated by eserine and pilocarpine, either alone or as an adjunct to iridectomy or sclerotomy, but in the acute and sub-acute forms an operation is urgently called for.

These drugs contract the pupil and relieve the obstructed drainage by dragging a portion of the iris away from the clogged "angle of filtration." Mydriatics like atropine, on the other hand, increase the tension by pushing the mass of the iris towards its periphery and so preventing drainage. *They should, therefore, never be prescribed in glaucoma.* Moreover, in old people, it is wise to determine the degree of ocular tension before prescribing mydriatics.

Still more important is it that the surgeon shall always be certain that the case of "iritis," for which he is prescribing atropine, is not one of acute or "congestive" glaucoma.

Ordinary care, combined with a knowledge of the symptoms and signs proper to both diseases, will prevent such a lamentable error.

LESSON IX.

OCULAR AFFECTIONS IN GENERAL DISEASES.

Manifestations of Syphilis, Rheumatism, and Other Diatheses—Muscæ Volitantes—Amblyopia—Toxic Amblyopia—Eye Symptoms in Tobacco and Alcohol Poisoning—Abscess of the Orbit—Graves' Disease—Progressive Locomotor Ataxia—Diphtheria—Bright's Disease — Migraine — Malaria — Reflex Neuroses—*Sympathetic Ophthalmia*—Penetrating Wounds of the Globe—Sympathetic Irritation and Inflammation—Treatment—Enucleation of the Eyeball.

Many scrofulous, syphilitic, tubercular, rheumatic, and gonorrhæal affections of the eye have been spoken of in these pages. In addition to these local manifestations of constitutional diatheses, there are certain signs and symptoms exhibited by the visual apparatus and complained of by the patient, whose origin lies remote from the organ of sight, that do not indicate organic ocular disease at all, or point only to a partial or secondary involvement of the eye. These facts continually teach us that a true specialism is that which regards the organism as a whole, and he who would successfully treat the diseases of one part of it must ever cast side glances at the condition of the other organs and tissues. Some of these ocular signs of remote disease will now be considered.

Muscæ Volitantes. Patients not infrequently complain of small bodies floating across the field of vision. They are especially noticeable when looking at the sky, or at some other bright background, when dark threads or spots dart like insects (*muscæ*) in front of the eye. These appearances are due to minute vitreous specks which almost every normal eye possesses. When the health is poor, the digestion bad, the circulation weak or, in woman, when the reproductive organs are diseased, these minute bodies in the vitreous humor may make their presence felt, and although they do not interfere with vision they yet give rise to a good deal of complaint. In addition to this condition, however, it must be remembered that the vitreous may become diseased and infiltrated with small opacities which not only lower the visual acuity but produce permanent *muscæ*. These floating bodies are readily detected by the examination with reflected light (20). The patient looks slowly in various directions and then at the mirror, the observer's gaze being directed at the pupil. The dark bodies rise and fall with the motion of the globe and are seen, especially when the pupil has been dilated, to move across the bright field of the "red reflex."

Pathological vitreous opacities are often the remains of effused blood or they may be produced by inflammation (hyalitis) of the corpus vitreum itself. They may also arise from injury or as a sequence of disease of the choroid and other deep structures of the eye.

Treatment is of little avail except in those cases

aggravated by a lowering of the systemic tone. General treatment in such cases often brings relief. Where the muscæ are manifest from uterine, hepatic, or other affections the duty of the physician is obvious.

AMBLYOPIA. This term belongs to the pre-ophthalmoscopic period of ophthalmology, but it still survives to describe certain diseases whose chief symptom is deficient vision without constant and evident organic lesion. Amblyopia is sometimes congenital and is then probably due to deficient development of some part of the optic tract. Or it may arise from simple non-use of the eye, as in squint (141). Hence the term *Amblyopia ex anopsiâ*. There is an important variety of this affection whose symptoms every practitioner is likely to meet with sooner or later, although it is by no means a common disease. I refer to the visual disturbances produced by certain (intoxicant) poisons, especially alcohol and tobacco.

TOXIC AMBLYOPIA. This disease is pathologically (Uhthoff) an interstitial neuritis, affecting preferably the macular fibres of the optic nerve and closely resembles those neuritic processes which one sees in the peripheral nerves of patients suffering from aggravated forms of chronic alcoholism. It is now generally admitted that, so far as the optic nerve is concerned, tobacco may also produce such a true *chronic retro-bulbar neuritis;* indeed most English observers claim that nicotine and not alcohol is responsible for the result. The association of the two poisons in bringing about the morbid changes is the commoner experience. The disease almost invariably attacks men above 40 years of age and affects both eyes.

The *symptoms* are characteristic and consist chiefly of a complaint that vision is cloudy or foggy. The usual story is that a "mist" appears to hang about objects upon which the patient fixes his gaze. Reading fine print becomes difficult or impossible; the vision is lowered, often markedly so ($\frac{20}{50}$ or $\frac{20}{40}$); there is a general atonic condition, some insomnia, anorexia and there may be a dull frontal pain. This complaint made by a constant smoker (or drinker) over 40 years of age should at once arouse the suspicions of the practitioner. An examination of the vitreous will be in order to make certain that the "fogginess" is not due to opacities in that medium, and the lens (100) to be sure that there is no commencing cataract. But the test *par excellence* is the examination of the color sense (14) or the field of vision for color.

Take a single thread of bright red Berlin wool and double it between the thumb and finger of the right hand, holding it so that only a quarter of an inch of the loop projects and is seen. Cover the patient's left eye with a bandage, and stand in front of him so that he can fix the examiner's right eye at a distance of a few feet. Now interpose the small piece of red wool. When placed directly between the two right eyes, the patient will not recognize it as red (or the red color will be indistinct) until it is moved two or three inches to the right, left, up and down from the line joining the two eyes. The same result will be obtained from *green* wool, but probably not from white, blue or yellow unless the disease is far advanced and the visual acuity lower than (12) $\frac{20}{200}$. In other words, the existence of a central scotoma *for red and green* has been demonstrated—the characteristic sign of toxic amblyopia.

Prognosis is favorable, unless the vision has been less than $\frac{20}{200}$ for a considerable time before applying for treatment. Taken in time, perfect recovery will follow an observance of the following rules: 1st, total and immediate abstinence from the use of tobacco and alcohol. 2d, Treatment of the general health, good food, etc. 3d, The administration of medium doses of iron and strychnia by the mouth, and single daily (if possible) hypodermic injections of strychnia. The dose of strychnia *et per orem et sub cutem* is to be slowly increased.

The diagnosis once correctly made, this routine treatment can be prescribed with the assurance that, if faithfully followed, a cure may be confidently expected in from one to three months. A guarded diagnosis will, of course, be given when the case is very chronic and vision is greatly lowered.

ORBITAL ABSCESS. Pus within the orbit may have various origins. (1) It may break through from the frontal sinus; (2) from a carious tooth; (3) from a tonsillar abscess: (4) it may be an extension of facial erysipelas; (5) it may follow an attack of cellulitis due to many causes, such as injury, etc.; (6) it may follow syphilitic and strumous orbital periostitis; (7) the abscess may be metastatic. In the acute forms of cellulitis and periostitis we notice that the ocular excursions are limited, the eye is more or less displaced, the conjunctiva is chemosed, while the lids are thickened, puffy and red. If the disease be near the margin of the cavity, pain will be severe and tenderness is usually well marked. If deep, there will be evening rise of temperature, chills, and the other symptoms of septic absorption. As it is difficult to say just where pus has formed, an exploring needle should be used if there is doubt. The abscess sooner or later points somewhere between the globe and the rim of the orbit, and should be opened through the conjunctiva with a long and narrow knife at as early a date as possible. The pus cavity should be thoroughly washed out and a drainage tube inserted. Free drainage must be maintained at all hazards and the possibility of cerebral complications not forgotten. If there be periostitis of specific origin, constitutional treatment should be given. When the disease lies deep and relief is not obtained early, the optic nerve may atrophy from pressure.

Fistulous openings, the result of a partially cured abscess, should be treated like fistula elsewhere.

EXOPHTHALMIC GOITRE. Graves' or Basedow's disease. In addition to the fast pulse, enlarged thyroid, nervousness, and some other symptoms of Graves' disease the ocular signs are very important. Ninety per cent. of cases occur in women. The *exophthalmos* (10) may be well or slightly marked. It may be so great that dryness and ulceration of the cornea results from exposure and, during sleeping hours, the eyelids are no longer able to cover the projecting globe.

Graefes' sign is almost characteristic of the disease and is not only one of the earliest but one of the most valuable guides to diagnosis. It results from the impairment of the consensual movement of the eyeball with the upper lid. In healthy persons the lid rises and falls (the palpebral edge always preserving the same relation to the upper corneal margin) as the globe rotates upwards or downwards. But in Graves' disease this relation is markedly altered. If the patient fix the observer's finger tip at two feet, while it is being slowly moved from above the patient's head downwards in front of his face, the lid will be seen to follow the eye for a short distance and then begin to lag behind until an abnormal portion of the sclerotic is visible between them.

Dalrymple's (Swanzy) *sign* is the staring appearance of the patient. The interpalpebral aperture is wider than normal and gives rise to this condition. Cocaine will produce a similar appearance; hence it is that some observers have thought that Dalrymple's sign is caused by a partial anæsthesia of the conjunctiva — a condition which is sometimes present.

Stellwag's sign is also valuable and fairly constant. It consists in the infrequency of winking. The act is also incomplete, for it will be noticed that the lid margins do not touch. Nictitation occurs once in from 40 to 70 seconds by the watch instead of every 20 or 30 seconds, as it should.

Prognosis. As Hulke says, in exophthalmic goitre we have the rare instance of a disease which runs an exceedingly chronic course (lasting for years) and yet tends, if life lasts, to get entirely well, and this although ten per cent. of the cases succumb to intercurrent diseases. So far as the eye symptoms are concerned the proptosis (123) does not often disappear, although it usually undergoes marked improvement. Tarsoraphy (47) may be performed if the cornea remains exposed during sleep.

PROGRESSIVE LOCOMOTOR ATAXIA. Tabes dorsalis. The eye symptoms of this serious disease are extremely important. *Vision* is one of the first senses impaired, and it may be lowered early in the disease and go rapidly on to blindness through atrophy of the optic nerve. As a rule, however, it either does not advance beyond a certain point, or its progress is very slow. Visual disturbances occur (Ross) in 30 per cent. of all cases. *Paralysis of the ocular muscles* is a common and early sign of tabes, and it is present in such a large percentage of cases that the occurrence of ocular pareses (ptosis, squint, etc.) in an adult should always give rise to suspicions of this disease.

The paralysis is at first partial and transitory, but as the spinal degeneration advances it becomes permanent and incurable. The patient complains, of course, of double vision (136). If it be affected at all *the pupil* is nearly always contracted. In Eulenburg's collection of 64 cases of tabes 28 had myosis (95). But the most important eye-sign of locomotor ataxy is the *Argyll-Robertson pupil;* that is, the pupil contracts to accomodation (95), but will not contract to light (8). Even when myosis is present the small pupil becomes still smaller when the patient directs his gaze from infinity to the near point.

DIPHTHERIA.—The poison of this disease gives rise, in a fair percentage of cases, to paralysis (cycloplegia), more or less complete, of the ciliary muscle. The accommodation is inactive, and the pupils are sluggish and dilated. In fact, the condition is much the same as if weak atropine drops had been instilled. If a child presents himself complaining of sudden inability to read or to see close at hand, but with good vision in the distance, and the use of a mydriatic is excluded, an examination will be very likely to disclose the fact that he has had diphtheria some four or six weeks previously. Such patients have a nasal intonation due to paralysis of the soft palate.

Prognosis is good. Even without treatment—which should be directed towards the general condition—the eye symptoms almost always slowly but entirely disappear. A suitable reading glass may be ordered while the paralysis of accommodation lasts.

BRIGHT'S DISEASE.— In the various nephritic affections included under this heading, vision is rarely disturbed until the disease is well established. The ocular lesion consists of a degenerative inflammation of the retina, and it is one of the gravest manifestations of Bright's disease. More than one-half of such patients die within a year, and many of them within a few weeks or months, after their sight is thus affected. Disturbances of vision due to this deep-seated inflammation occur in about 25 per cent. of all cases of albuminuria with organic disease.

The *prognosis* is, of course, unfavorable, although some cases, associated with puerperal albuminuria and with scarlatina, get well. In these instances, however, it happens that perfect vision is never recovered, because irreparable damage has meantime been done to the percipient elements of the retina or to some of the optic nerve fibres.

MIGRAINE, *megrim*, or *sick headache*, is frequently accompanied by temporary disorders of sight. Either before or during an attack of this distressing form of neuralgia a peculiarly shaped cloud appears directly in front of the eyes. It begins as a dark, central scotoma having a bright colored margin, serrated like a line of fortification, and called, for this reason, "bastion scotoma." This spreads until the whole field of vision is obscured, and lasts but a short time, after which sight is as perfect as before the attack. Often persons suffering from sick headache notice muscæ (117) and fogginess only, instead of the well defined scotoma just described. It has also been observed that persons subject to migraine always have refractive errors (R. 50).

Prognosis is favorable.

Treatment will be directed to the cause of the headache. Amyl nitrite (Noyes), in from 3- to 5-minim doses, will relieve the severe and prolonged attacks. Attention should be paid to the refractive condition of the eye, and when ametropia (C. 21) is present proper correcting glasses should be prescribed.

MALARIA has been blamed for many forms of ophthalmic disease, such as chronic conjunctivitis, keratitis, etc.; but beyond finding circumcorneal infection (83) and a few cases of iritis, I cannot honestly say that I have seen many ocular affections of well-defined malarial origin. When such do occur, the treatment by anti-periodics is not so successful as in other forms of malaria.

REFLEX NEUROSES.—Roosa (50) has spoken of the many nervous affections that directly result from defects of accommodation and errors of refraction, and has shown how frequently the oculist is called upon to treat nervous disorders having their origin in optical deficiencies. In addition to these disturbances of vision and of the ocular circulation, actual inflammation of the conjunctiva, iris, etc., are produced by sympathy with diseased organs more or less distant from the eye. *Nasal diseases* are among these. We have already seen (50) how they may extend along the nasal duct to the lachrymal and conjunctival sacs. But in addition to this, pain in the eye, conjunctival and ciliary hyperæmia, epiphora (49), and occasional dimness of vision may be produced as purely reflex phenomena from such nasal troubles as stenosis from ecchondroses, hypertrophies, polypi, bony spines on the septum, empyæma of the maxillary sinus, and so on.*

Much the same train of symptoms has resulted from *decayed teeth*. *Uterine diseases*, as well as *venereal excesses*, are sometimes the sources of similar symptoms.

SYMPATHETIC OPHTHALMIA. By this term we mean the involvement of an eye by disease which has spread from the other eye by way of the optic nerve. The first ("exciter"), or "exciting" eye is, almost without exception, an injured eye, and the second ("sympathizer"), or "sympathizing" eye, becomes, with almost equal uniformity, the subject of an iridochoroiditis or of an inflammation of the whole uveal tract—iris, choroid, and ciliary body. The most dangerous wounds, so far as concerns the liability to sympathetic ophthalmitis, are those that penetrate the region (which Nettleship calls the "dangerous zone") corresponding to the ciliary body. This area, 4 or 5 mm. in width, extends around and a few millimetres outside of the sclero-corneal junction. Foreign

---

*See Boerne Bettman's article, Journal American Medical Association, May 7, 1887.

bodies lodged in the interior of the bulb, as well as perforating ulcers of the cornea, may also light up the disease.

Deutschmann and Gifford have demonstrated to a certainty the fact that germs from the inflamed and "exciting" eye are carried along the optic nerve to the chiasma, and thence to the uveal tract of the sound eye, where they set up an inflammation of a sero-plastic type. From these circumstances the disease has been called *ophthalmia migratoria*.

SYMPATHETIC IRRITATION.—This may be a premonitory stage of the succeeding ophthalmitis, or it may not proceed further. The chief sign of it is tenderness on pressure over the ciliary region—the patient draws his head away when the eye is pressed upon (9). There is also slight photophobia, and some ciliary injection. Usually there is no pain whatever. The stealthy setting in of these symptoms on the sound side, after a penetrating wound of the other eye, may well cause the surgeon some anxiety. He should be on the lookout for them at almost any date subsequent to three weeks after such a traumatic lesion. Sympathetic ophthalmia has, indeed, been observed as early as two weeks and as late as 20 years after injuries. ' In other words, an eye containing a foreign body, or one which has been the subject of traumatic cyclitis, is a dangerous eye, and liable at any time to bring ruin upon itself or its fellow organ.

After the slight warnings just spoken of, definite

changes show themselves in the sympathizing eye. To the tenderness are added occasional slight pains, more photophobia and more lachrymation, while vision becomes sensibly impaired. One soon notices, besides the pericorneal injection, a serous iritis with keratitis punctata (69) and a deep anterior chamber (7). Then plastic deposits take place in the ciliary body, the vitreous becomes cloudy, cataract invades the lens, the nutrition of the whole eye is interfered with, and after a longer or shorter period of suffering marked, it may be, by teasing pains in the eye, atrophy of the bulb (*phthisis bulbi*) results with total loss of vision. If relief is given, the eye may partially or wholly recover; but, whether it does or not, the course of the disease is always very chronic, very wearisome and very variable. It may, in fact, happen that months, or even years, after an attack of sympathetic inflammation, the injured eye has better vision than its fellow.

Treatment.—The treatment of migratory ophthalmia requires special care and special knowledge. The conduct of a case in which signs of this dreaded disease appear should not be lightly undertaken by a non-specialist. The most important rules for him to remember are those which refer to the removal of the "exciting" eye. When an eye is so injured that no useful vision remains or none can be preserved, it should be enucleated (132). When this is not done the case ought to be constantly watched until the injured eye has entirely healed and becomes "quiet," or until symptoms of "sympathy" show themselves in the other eye. An attempt should be made to secure the former result by bringing the edges of the wound into apposition. Sutures may be applied if in the sclera. Prolapsed iris and vitreous must be cut off with the scissors and the stump of the former replaced (95). Blood clots must be removed, the conjunctival sac thoroughly irrigated and the eye carefully disinfected (31); in fact, the wound and its surroundings must be made and kept as surgically clean as possible. The dressings (antiseptic gauze is the best) should be changed frequently, if there be any discharge or pain. This line of treatment must be persevered in until healing has taken place.

In the event of sympathetic symptoms appearing, and vision is fairly good in the injured eye, the problem of treatment is such a difficult one that want of space prohibits its discussion here. In addition to the hints already thrown out I would refer the reader to the excellent rules laid down by Swanzy (Diseases of the Eye, p. 230). When the practitioner or his patient cannot obtain competent advice during the varying phases of this troublesome affection it is best to sacrifice the injured eye, even when its vision is fair, on the first approach of sympathetic "irritation," that, happily, further advance of the disease may be stayed. At the same time it is right to remember, and wise to warn the patient, that even after the exciting eye has been excised two weeks must pass before one can feel certain that ophthalmitis may not develop in the sound eye. In other words morbific germs from the injured eye may have been *on their way* to or have already reached the other eye before the exciting eye was removed

When ophthalmia migratoria has set in and the vision in the injured eye is fairly good the usual plan among oculists is not to excise the offending organ (in a person who can wait) but to fight the disease with appropriate remedies, because, as before stated, it may happen that, after months of patient nursing, vision in the exciting eye is better than that of the uninjured eye.

ENUCLEATION or excision of the eye is usually performed with a strong pair of scissors (Fig. 30), curved on the flat, a pair of fixation forceps (Fig. 16) and a strabismus hook (Fig. 32). A strong solution of

FIG. 30.

cocaine is relied upon by some operators as an anæsthetic. It is applied to the cornea and injected behind the bulb as soon as the capsule of Tenon (144) is opened, but when the eye is inflamed very little of the solution is absorbed and then it is best to give ether or chloroform. A speculum is introduced and the eye being fixed the conjunctiva is cut through with the scissors all round the cornea. Each rectus tendon is now severed on a strabismus hook close to the globe. Now separate the branches of the speculum and the eyeball will start forward. Space is given, in this way, to cut carefully through the remaining muscles, faschia, and other attachments, always keeping close to the globe. Last of all the optic nerve is divided and the eyeball comes away. Hæmorrhage may be free but it is readily controlled by plugging the orbit. The conjunctiva is left to itself; a simple boracic lotion is used to bathe the parts and, in most cases, simple gauze dressing applied over the lids is all that is required. In from 3 to 6 weeks time, when all irritation and discharge have subsided, an artificial eye may be worn.

LESSON X.
PARALYSIS, SQUINT AND OTHER MUSCULAR TROUBLES.

The Physiology of the Subject—The Nerve Supply—*Ocular Paralyses*—Their Symptoms—Paralysis of the Sixth Nerve—Paralysis of the Fourth Nerve—Oculo-motor Paralysis—Ophthalmoplegia—Causes and Treatment of Paralysis—*Strabismus or Squint*—Convergent and Divergent Squint—The Measurement of Squint—Treatment—Operations for Strabismus — Tenotomy — Advancement.

The centre about which the eyeball rotates is situated in the line of its visual axis about 14 mm. behind the cornea. Three pairs of muscles move it in various directions. The separate action of the *rectus externus* is to rotate the eye outwards, of the *rectus internus* to move it *inwards*, while more complicated movements in various directions are effected by the combined action of these with the *superior* and *inferior oblique* muscles. The fourth nerve supplies the superior oblique, the external rectus is supplied by the sixth nerve while the other ocular muscles (including the *levator palpebræ superioris*, the sphincter pupillæ and the ciliary muscle) are under the influence of the oculo-motorius – the third cerebral nerve.

When a person, with erect head, looks at a distant object directly in front of him and in the horizontal plane, head and eyeballs are said to be in the *primary position.* This is accepted as a sort of standard with which to compare all other positions of the globe.

Abnormal attitudes of the eyeball are taken in cases of *heterophoria*—muscular insufficiencies – (R. 45) (8), *ocular paralysis* and *strabismus* or squint.

Diagnosis. The detection of insufficiencies has already been spoken of (11). When affected by either squint or paralysis, both eyes are not directed towards the same quarter in all positions of both globes. One disease may easily be differentiated from the other by the simple expedient of testing the excursion (11) of each eyeball in all directions. Examined separately they will be found to have a normal excursion in squint while restricted movement in one or more directions can be detected when a muscle (or muscles) is affected by paralysis.

PARALYSIS OF THE EYE MUSCLES. Although it is usual to speak of paralysis of the eye muscles yet, for clinical reasons, it is advisable not to forget their nerve supply (134). For, as a matter of fact, it is the nervous function that is disturbed or abolished, and if one recollect the ocular innervation paralytic diseases of the muscles resolve themselves naturally into well defined clinical groups, as we shall see. It is mainly for purposes of diagnosis—when one wishes to discover what particular muscle or muscles are involved—that prominence is given to the loss of *muscular* function.

There are certain symptoms common to all forms

of paralysis. The most important of these is *diplopia* —the patient sees double. This occurs in every instance where vision in both eyes is good, and is due to the fact that images of objects do not fall on *corresponding* parts of both retinæ. It is by the relation of these double images—a somewhat difficult subject for the student—that most authors seek to indicate the particular muscle affected. *Giddiness* and even *nausea*—the nervous effects of the diplopia and of the false projection—as well as indistinct vision are symptoms frequently complained of. *Headache* is not uncommon. The patient, to avoid the annoyance of double vision, will usually close one eye or turn his head towards the paralyzed muscle. This sign often indicates which muscle is affected.

PARALYSIS OF THE EXTERNAL RECTUS or of the sixth nerve. This is easily recognized and is probably the commonest form of the ocular pareses. The patient has double vision and the other symptoms mentioned, and when the head is in the primary position (134) the eyes converge. The diagnosis between this disease and convergent squint may be easily made by the method just referred to (135).

PARALYSIS OF THE SUPERIOR OBLIQUE or of the fourth nerve. The ocular excursion downwards and inwards is defective in this paralysis. In the field above the horizon there is single vision but below it diplopia. In looking downward at the sidewalk objects are seen displaced and distorted so that walking is difficult or impossible. The lower limbs of people assume a mixed and multiplied appearance, while their heads and faces are natural. It is difficult to measure accurately the height of a step one is about to put the foot upon, etc. Much the same symptoms are present in those cases where the inferior rectus *alone* is involved.

OCULO-MOTOR, or third nerve, paralysis. Any one, two, three, four, five, six or all (134) of the muscles supplied by the third nerve may be paralyzed. Usually, however, there is ptosis (45), from paralysis of the levator palpebræ, with mydriasis and loss of accommodation due to involvement of the sphincter pupillæ and ciliary muscle. The paralysis of other muscles can be made out by the loss of motion proper to each. It must not be supposed that it is always easy or possible to say just what muscle or muscles are affected on account of the *secondary* contractions and deviations that occur in both eyes, owing to efforts to obtain binocular vision.

OPHTHALMOPLEGIA INTERNA AND EXTERNA. The former term is applied to paralysis of the sphincter pupillæ and ciliary muscles when it occurs alone. In the latter also called from its origin, *nuclear* paralysis, all or most of the external muscles are affected to the exclusion of the ciliary muscle and pupillary sphincter.

Causes of paralysis are chiefly rheumatic or syphilitic affections, either of the nerves themselves in their course from the brain, or of their nuclei. Organic deposits in the bony canals along which most of the cerebral nerves run, or exostoses from their walls, as well as growths from the neurilemma, may exert pressure sufficient to bring about a temporary abolition or a total loss of their function. The reabsorption of these growths or deposits may result in a cure unless too great damage has been done to the nervous elements. Where one nerve alone is affected the cause is probably a peripheral one, while nuclear paralysis is to be suspected if more than one nerve suffers. Von Græfe's test must be borne in mind, viz.: when fusion of the double images by the use of prisms is easy the lesion is probably peripheral, but when it is difficult to obtain and retain single vision the paralysis is due to spinal or cerebral disease. Although syphilis and rheumatism play a very important role in the causation of these pareses it is sometimes difficult to demonstrate their presence. A few cases, however, result uniformly from one cause; paralysis of the external rectus, for example, almost invariably occurs in rheumatic subjects. *Diphtheria* sometimes produces orbital paralysis and, as before mentioned (125) is a cause of cycloplegia with dilated pupil (iridoplegia). Reference has already been made to the frequency of these paralyses in *locomotor ataxia* (124). Paralysis of the external rectus is not unusual in diabetes.

Prognosis. Diphtheritic paralysis and the *prim-*

ary paralysis of tabes almost invariably disappear. So do most of those that depend upon peripheral causes. If of central origin many syphilitic cases get well, but some do not. For obvious reasons the later tabetic pareses remain, as well as many others of central origin.

Treatment. It is justifiable to cover the affected eye with a shield so as to guard against the troubles of diplopia. Specific treatment will be given when it is indicated, and even when there is no definite history of syphilis potassic iodide, given in gradually increasing doses until 30 or 40 grs. are taken three times daily, may be continued for several weeks or months. Cupping the temple in the early stages and the employment of the constant electric current are remedies of extreme value. Cocainize the eye and place the negative pole (a small sponge) between the lids directly over the paralyzed muscle. The positive pole may be applied to the neck. This can be kept up for three or four minutes at a time and is a better plan than the usual application of a larger sponge to the closed lids. Passive motion (Michel) may be applied in this as in other forms of paralysis. Under cocaine the insertion of the muscle is seized with fixation forceps and the eyeball drawn or pushed in the direction of its contraction and back again. This is to be done once a day (or oftener) for a minute at a time. Surgical interference, as in strabismus (143), may be resorted to in long-standing cases when medical measures have failed to restore the lost muscular function.

STRABISMUS or SQUINT. This affection is sometimes called "concomitant" squint because although the relation of the visual axes is not a normal one it is a *constant* relation—one eye moves about when the other does. In "paralytic" squint this is not the case. Much confusion arises from the calling of paralytic diseases "squint" and it would be better to confine that term to the conditions about to be described.

The two most important and by far the commonest varieties of this disease are *convergent* and *divergent strabismus*. In the former case, when one eye fixes an object the other converges or turns in more than it should; in the latter instance the non-fixing eye diverges or turns out. Usually one eye does the "fixing" (and seeing) while the other squints. This is called *constant* or unilateral strabismus. Sometimes (and then both eyes have the same visual acuity) it seems to be a matter of indifference to the patient with which eye he fixes and which eye squints; sometimes it is one, sometimes the other. These form an important class of *alternating* squints. Other cases squint occasionally only—the so-called *periodic* squint.

It is easy enough to detect squint if the eye-covering test (11) be applied, but apart from tests most squinting eyes manifest themselves if the patient be directed to look first at a near point—say 30 cm. in front of his nose—and then at some distant object

Causes.—The causation of squint is wrapped in mystery notwithstanding all the investigations of the subject and all that has been written about it. Some very pretty theories have been advanced to explain all the facts, but none has yet done so satisfactorily. It may be said, speaking in general terms, that while the optical centres of most (not all) individuals *prefer* binocular vision they *insist upon* clear images of objects. So that, if one optic nerve receives and conveys to the cerebral centres the sensation of a blurred image and the other, at the same time, transmits a clear or clearer image the fiat goes forth to suppress the less distinct image. This is done by sacrificing binocular vision, making the eye turn in or out according as convergent or divergent power predominates. Now, the interni muscles being strongest in hypermetropes and emmetropes we find that these persons are almost always subject to convergent squint. Myopes, on the other hand, have relatively strong externi and weak interni; hence the strabismus in myopia is of the divergent kind.

Strange to say there is no double vision in strabismus, since the brain suppresses the indistinct image of the squinting eye—just as when "in a brown study" one's retina does not perceive surrounding objects.

Convergent strabismus usually sets in between the ages of one and five when the child has begun to use its interni muscles for convergence. The great majority of us are born hypermetropic (R. 36) and that is, perhaps, the reason why most squinting children are affected by convergent squint. Divergent strabismus, on the other hand, is less frequent in this country owing to the comparative rarity of myopia. Short-sightedness is a disease of adolescence; hence divergent squint develops later in life than the convergent variety.

FIG. 31.

The *degree* of squint is measured by the *strabometer* and other instruments. The former is pictured in Fig. 31. The patient is in the primary position (134) and the instrument is placed along the lower orbital margin of the squinting eye. The centre of the pupil will now be found opposite a number which indicates in lines (or millimetres) the amount of deviation. This must be done both for near and distant fixation.

Treatment.—The patient must, first of all, be given *full* correction of all refractive errors. Weak atropine drops are also prescribed for a few weeks and the glasses worn constantly. In a fair percentage of hypermetropic cases this alone will bring about a complete cure in the course of several months or a year. If, after this trial, little or no improvement results an operation is indicated.

Assuming that glasses are worn and atropine used, the following rough rules will serve to indicate the date, the amount and the kind of operative interference necessary: 1. If possible the operation should be done on the squinting eye. 2. Use cocaine for tenotomies and chloroform for advancements. 3. The more the tendon is loosened from its connective tissue bed the greater the effect of the tenotomy. 4. In *convergent* squint, where the strabismus is not more than one line, a free tenotomy of one internal rectus may be sufficient. When the deviation amounts to two lines both interni should be divided. More than that calls for section of one internus with advancement of the externus of the same eye. 5. Slight degrees of *divergent* squint call for a tenotomy of the external rectus. Marked deviations will need, in addition, advancement of the internus muscle. 6. When the squinting eye is amblyopic (119) or when from other causes, such as corneal opacities, cataract, etc., its vision is but slightly or not at all improved by glasses, the spectacle and atropine treatment exerts no influence upon the squint and for the sake of appearances (cosmetic effect) the operation should be proceded with at once.

Tenotomy of a muscle is done in the following fashion: The eye having been well cocainized, is rolled over by fixation forceps (Fig. 14) to the side opposite to that on which the operation is to be done, and is retained in position by the assistant. The conjunctiva and sub-conjunctival tissue immediately over the tendinous insertion are caught up by another pair of forceps, and a fold of mucous membrane cut through by a pair of straight scissors at the lower edge of the tendon. The points of the scissors are now passed into the aperture, Tenon's capsule is opened, and the tissues lying over and on both sides of the muscle are undermined as much as necessary. Next, the strabismus hook (Fig. 32) is slipped into

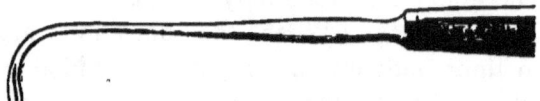

FIG. 32.

the opening, and with a half turn is made to pass *under* the tendon. This last manœuver requires some practice. Be sure that the hook point is applied to the globe, and that it is far enough back before rotating its point underneath the muscle. On drawing it forward there should be a feeling of resistance, and one should make certain that the point presents free of all the tissues on the *other side* of the muscle. Now divide the tendon, put somewhat on the stretch, between the hook and the eyeball, as close to the latter as possible. The hook will give way and come forward to the corneal margin, showing that the tendon, and not merely some connective tissue fibres, have been severed. With the curve of the hook sweep round on both sides of the cut tendon so as to catch up and divide all remaining fibres. If the conjunctival opening is small no suture is necessary.

Advancement of a muscle is done under ether or chloroform. There are many methods, but that of Schweigger has my preference. The conjunctiva is well divided over the muscle, the latter being thoroughly exposed and well cleaned of connective tissue. The mucous membrane is now extensively undermined on all sides, quite up to the corneal margin and along one-third of its circumference. Two strabismus hooks are passed underneath the muscle (one from each side), or an advancement forceps (Prince's is best: see Fig. 33) is made to grasp the

Fig. 33.

muscular body so as to hold it steady and away from its bed. A double-needled piece of catgut is passed through the centre of the muscle and is tied firmly at its edge *below*. The same thing is done *above*. The muscle is next cut off close to the sutures on its bulbar side, and the needles of each suture are directed underneath and through the undermined conjunctiva (two above and two below) well forward and close to the cornea. The muscle may now be drawn towards the corneal margin until the eyeball is made to assume the desired position. Each half suture is now tied to its fellow over the intervening conjunctiva. The original opening in the mucous membrane is stitched together by fine sutures. There may be some reaction following this operation requiring the frequent application of hot fomentations, but if proper precautions (31) have been taken this is unusual.

INDEX.

The student is advised to read over the Index carefully, looking up the references to those subjects with which he is not familiar.

A.

	PAGE.
Abscess of cornea	68
lachrymal	51
orbital	121
Acuity, visual	12
tests of	13
Acid boric, use of	26
Advancement of ocular muscles	145
Albinism	79
Alcoholic amblyopia	120
Amblyopia	119
toxic	119
Anel's syringe	54
Arcus senilis	56
Aphatia	98
Astigmatism	74, 75
Atrophy of optic nerve	95
Atropine	22, 89
irritation	32

B.

Bandages for eye	29
Basedow's disease	123
" Black eye "	36
Blepharitis marginalis	34, 50, 102

	PAGE.
Blood-vessels, anterior ciliary	10
of conjunctiva	6
episcleral	11, 83, 114
Bowman's probes	53, 55
Bright's disease, eye symptoms in	126

C.

Calomel in eye diseases	27
Canaliculi	4
Canaliculus, slitting of	52
Canthoplasty	41
Cartilage, tarsal, or tarsus	6
Caruncle	4
Cataract, varieties of	97, 99, 109, 110
operations for	105, 108, 110
Cautery applications	31, 72
Chalazion or tarsal cyst	38
Chamber, anterior	7, 114
Cilia, normal	4
forceps	35, 42
Cocaine, action and use of	24
Color perception, test of	14, 120
Coloboma of iris	79
Conjunctiva, normal	5
Copper, sulphate of	27
Coquilles	30
Cornea, abscess of	68
foreign bodies in	65
normal	7
speed	66, 73
tattooing the	77
ulcers of	57, 59, 69
Cycloplegia	138
Cystotome	104

D.

	PAGE.
Dacryocystitis...	49
Diplopia...	136
Diphtheria, eye signs in........................ 95, 125,	138
Discs, use of ophthalmic......................................	26
Discission for cataract........................... 110,	111
needle...	110
Dislocation of lens..	97
Distichiasis...	40
Duboisia, action of.. 23,	89

E.

Ecchymosis of lids...	36
Ectropion...	44
Entropion...	40
Holz's operation for...........................	42
Enucleation of eyeball................................... 77,	132
Epilation..	42
Epiphora..	49
causes of...	49
Episcleritis...	77
Eserine, action of....................................... 23,	116
Evisceration of globular contents........................	77
Excursion of eyeball...................................... 11,	135
Eye, inspection of normal..................................	3
"strain".. 36,	39
Eyes of children, examining................................	14
examination of diseased.................................	16
Eyelids, normal..	4

F.

Facets of cornea...	74
Forceps, cilia...	42
fixation..	92
iris...	94
Foreign bodies in cornea.....................................	65

G.

	Page.
Glands, meibomian	5
Glaucoma	95, 112
Goggles and *coquilles*	30
Goitre, exophthalmic	122
Graefe's cataract knife	92
Graves' disease	125

H.

Herpes of cornea	37, 74
lids	37
Heterophoria	135
Heterophthalmos	79
Hippus	96
Homatropine, action of	23, 98
Hordeolum	37
Holz's operation for entropion	42
Humor, aqueous	7
Hyalitis	118
Hyphœma	80
Hypopyon	71

I.

Inspection of normal eye	3
Iridectomy	77, 91, 115
optical	93, 110
Iridodialysis	80
Iridodonesis	97
Iridoplegia	138
Iris, normal	7, 82
coloboma of	79
forceps	94
repositor	95
scissors	94
Iritis, various forms of	80, 89, 91, 108, 127, 130
recurrent	94

K.

	Page.
Keratitis	56
interstitial	66
parenchymatous	66
phlyctenular	57
punctate	69, 130
Keratome	92, 93

L.

Lachrymal abscess	51
probes	53
sac	5
Lagophthalmos	47
Lapis divinus	28
Lead acetate, use of	28
Lens, absence of	98
crystalline	97
dislocation of	97, 112
Leucoma	75
adherens	72
Lids, ecchymosis of	36
examination of	5
Lithiasis, meibomian	34
Locomotor ataxia	124

M.

Macula of cornea	75
Malarial affections of eye	127
Meyer's syringe	54
Migraine, eye affections in	126
Migratory ophthalmia	130
Mucocele	51, 102
Muscæ volitantes	117, 127
Muscles, action of ocular	134
enervation of	134
equilibrium of	11
paralysis of ocular	135

N.

	Page
Nebula of cornea	75
Needling for cataract	110, 111
Nystagmus	80, 96

O.

Oblique illumination	9
Onyx	71
Opacities of cornea	75, 102
Operations, preparing for	31
Orbital abscess	121
Ophthalmoplegia	137

P.

Pagenstecher's ointment	26
Paracentesis corneæ	73
Paralysis of external rectus	136
third nerve	95
Phlyctenular keratitis	57
Photophobia	64, 67, 81, 86
Phthisis bulbi	130
Pilocarpine, use of	25
Pinguecula	33
Polyopia	100
Primary position	134
Probes, lachrymal	53
Projection of light	103
Pterygium	33
Ptos s	45
Puncta	4
Punctum, eyelash in	38
Pupils, normal	8
Pupillary reaction	8

R.

	Page.
"Red reflex"............................20, 101,	118
Reflex symptoms in eye diseases....................	127
Rodent ulcer..	40

S.

Sæmisch's section.....................................	73
"Salmon patch"......................................	67
Schlemm, canal of.......................8, 112,	115
Scissors, enucleation.................................	132
iris..	94
Sclera or sclerotic coat...............................	7
Scleritis..	77
Sclerotomy...	115
Scotoma.......................................121,	126
Shades for the eye....................................	30
Silver nitrate, use of.................................	28
Skin, palpebral.......................................	4
Snellen's lid-clamp...................................	39
Speculum for lids....................................	92
Speed, corneal...................................66,	73
Staphyloma anterius............................74,	76
Stenopaic glasses.....................................	80
Squint, varieties of.......................135, 140,	143
Strabismus or squint.....................135, 140,	143
hook..	144
Strabometer..	143
Stye..	37
Symblepharon..	48
Sympathetic ophthalmia..............................	128
irritation.....................................129,	131
Synechia anterior.....................................	72
posterior.................................87, 88,	112

T.

	PAGE.
Tabes dorsalis	124
Tarsoraphy	47
Tatooing the cornea	77
Tenotomy of ocular muscles	143
Tension of globe	9, 113, 114
Testing drum	104
Test types of Jäger and Snellen	13
Tobacco amblyopia	119
Trichcheasis	40

U.

Ulcer of cornea..................57, 59, 69, 74

V.

Vitreous opacities.................. 118

X.

Xanthelasma.................. 40

Convenient Preparations for Surgeons

WE supply Antiseptic Liquid, Tablets of Bichloride of Mercury, for easily preparing solutions of any desired strength; Labbaraque's Solution, Solution Aluminium Acetate, Sulphur Bricks.

We furnish Cocaine in the following packages: Cocaine alkaloid, pure in crystals; Cocaine citrate, 4 per cent. solution; Cocaine hodrobromate, pure in crystals; Cocaine hydrobromate, 4 per cent. solution; Cocaine muriate, pure in crystals; Cocaine muriate, 2 per cent. solution; Cocaine muriate, 4 per cent. solution; Cocaine oleate, containing 5 per cent. of the alkaloid; Cocaine salicylate, 4 per cent. solution; Cocainized oil, 5 per cent.

Cascara Cordial and Glycerin Suppositories are eligible and satisfactory laxatives after operations.

Mosquera's Beef-Meal and Beef-Cacao are concentrated foods, highly nutritious, perfectly palatable, and may be administered in a variety of forms. They constitute ideal foods for those enfeebled by operative procedures.

Pepsin Cordial is an agreeable and efficient digestive and tonic.

Descriptive literature of our products sent to physicians on request.

PARKE, DAVIS & CO.,
DETROIT and NEW YORK.

BULLETIN of PUBLICATIONS
— OF —
GEORGE S. DAVIS, Publisher.

THE THERAPEUTIC GAZETTE.
A Monthly Journal of Physiological and Clinical Therapeutics.
EDITED BY
ROBERT MEADE SMITH, M. D.
SUBSCRIPTION PRICE, $2.00 PER YEAR.

THE INDEX MEDICUS.
A Monthly Classified Record of the Current Medical Literature of the World.
COMPILED UNDER THE DIRECTION OF
DR. JOHN S. BILLINGS, Surgeon U. S. A.,
and DR. ROBERT FLETCHER, M. R. C. S., Eng.
SUBSCRIPTION PRICE, $10.00 PER YEAR.

THE AMERICAN LANCET.
EDITED BY
LEARTUS CONNOR, M. D.
A MONTHLY JOURNAL DEVOTED TO REGULAR MEDICINE.
SUBSCRIPTION PRICE, $2.00 PER YEAR.

THE MEDICAL AGE.
EDITED BY
B. W. PALMER, A. M., M. D.
A Semi-Monthly Journal of Practical Medicine and Medical News.
SUBSCRIPTION PRICE, $1.00 PER YEAR.

THE WESTERN MEDICAL REPORTER.
EDITED BY
J. E. HARPER, A. M., M. D.
A MONTHLY EPITOME OF MEDICAL PROGRESS.
SUBSCRIPTION PRICE, $1.00 PER YEAR.

THE BULLETIN OF PHARMACY.
EDITED BY
B. W. PALMER, A. M., M. D.
A Monthly Exponent of Pharmaceutical Progress and News.
SUBSCRIPTION PRICE, $1.00 A YEAR.

New subscribers taking more than one journal, and accompanying subscription by remittance, are entitled to the following special rates.
GAZETTE and AGE, $2.50; GAZETTE, AGE and LANCET, $4.00; LANCET and AGE, $2.50; WESTERN MEDICAL REPORTER or BULLETIN with any of the above at 20 per cent. less than regular rates.
Combined, these journals furnish a complete working library of current medical literature. All the medical news, and full reports of medical progress.

GEO. S. DAVIS, Publisher, Detroit, Mich.

IN EXPLANATION
OF
The Physicians' Leisure Library.

We have made a new departure in the publication of medical books. As you no doubt know, many of the large treatises published, which sell for four or five or more dollars, contain much irrelevant matter of no practical value to the physician, and their high price makes it often impossible for the average practitioner to purchase anything like a complete library.

Believing that short practical treatises, prepared by well known authors, containing the gist of what they had to say regarding the treatment of diseases commonly met with, and of which they had made a special study, sold at a small price, would be welcomed by the majority of the profession, we have arranged for the publication of such a series, calling it **The Physicians' Leisure Library.**

This series has met with the approval and appreciation of the medical profession, and we shall continue to issue in it books by eminent authors of this country and Europe, covering the best modern treatment of prevalent diseases.

The series will certainly afford practitioners and students an opportunity never before presented for obtaining a working library of books by the best authors at a price which places them within the reach of all. The books are amply illustrated, and issued in attractive form.

They may be had bound, either in durable paper covers at **25 Cts.** per copy, or in cloth at **50 Cts.** per copy. Complete series of 12 books in sets as announced, at **$2.50,** in paper, or cloth at **$5.00,** postage prepaid. See complete list.

PHYSICIANS' LEISURE LIBRARY

PRICE: PAPER, 25 CTS. PER COPY, $2.50 PER SET; CLOTH, 50 CTS. PER COPY, $5.00 PER SET.

SERIES I.

Inhalers, Inhalations and Inhalants.
   By Beverley Robinson, M. D.
The Use of Electricity in the Removal of Superfluous Hair and the Treatment of Various Facial Blemishes.
   By Geo. Henry Fox, M. D.
New Medications, Vol. I.
   By Dujardin-Beaumetz, M. D.
New Medications, Vol. II.
   By Dujardin-Beaumetz, M. D.
The Modern Treatment of Ear Diseases.
   By Samuel Sexton, M. D.
The Modern Treatment of Eczema.
   By Henry G. Piffard, M. D.

Antiseptic Midwifery.
   By Henry J. Garrigues, M. D.
On the Determination of the Necessity for Wearing Glasses.
   By D. B. St. John Roosa, M. D.
The Physiological, Pathological and Therapeutic Effects of Compressed Air.
   By Andrew H. Smith, M. D.
Granular Lids and Contagious Ophthalmia.
   By W. F. Mittendorf, M. D.
Practical Bacteriology.
   By Thomas E. Satterthwaite, M. D.
Pregnancy, Parturition, the Puerperal State and their Complications.
   By Paul F. Mundé, M. D.

SERIES II.

The Diagnosis and Treatment of Haemorrhoids.
  By Chas. B. Kelsey, M. D.
Diseases of the Heart, Vol. I.
  By Dujardin-Beaumetz, M. D.
Diseases of the Heart, Vol. II.
  By Dujardin-Beaumetz M. D.
The Modern Treatment of Diarrhoea and Dysentery.
  By A. B. Palmer, M. D.
Intestinal Diseases of Children, Vol. I.
  By A. Jacobi, M. D.
Intestinal Diseases of Children, Vol. II.
  By A. Jacobi, M. D.

The Modern Treatment of Headaches.
  By Allan McLane Hamilton, M. D.
The Modern Treatment of Pleurisy and Pneumonia.
  By G. M. Garland, M. D.
Diseases of the Male Urethra.
  By Fessenden N. Otis, M. D.
The Disorders of Menstruation.
  By Edward W. Jenks, M. D.
The Infectious Diseases, Vol. I.
  By Karl Liebermeister.
The Infectious Diseases, Vol. II.
  By Karl Liebermeister.

SERIES III.

Abdominal Surgery.
  By Hal C. Wyman, M. D.
Diseases of the Liver.
  By Dujardin-Beaumetz, M. D.
Hysteria and Epilepsy.
  By J. Leonard Corning, M. D.
Diseases of the Kidney.
  By Dujardin-Beaumetz, M. D.
The Theory and Practice of the Ophthalmoscope.
  By J. Herbert Claiborne, Jr., M. D.
Modern Treatment of Bright's Disease.
  By Alfred L. Loomis, M. D.

Clinical Lectures on Certain Diseases of Nervous System.
  By Prof. J. M. Charcot, M. D.
The Radical Cure of Hernia.
  By Henry O. Marcy, A. M., M. D., L. L. D.
Spinal Irritation.
  By William A. Hammond, M. D.
Dyspepsia.
  By Frank Woodbury, M. D.
The Treatment of the Morphia Habit.
  By Erlenmeyer
The Etiology, Diagnosis and Therapy of Tuberculosis.
  By Prof. H. von Ziemssen.

SERIES IV.

Nervous Syphilis.
  By H. C. Wood, M. D.
Education and Culture as correlated to the Health and Diseases of Women.
  By A. J. C. Skene, M. D.
Diabetes.
  By A. H. Smith, M D.
A Treatise on Fractures.
  By Armand Després, M. D.
Some Major and Minor Fallacies concerning Syphilis.
  By E. L. Keyes, M .D.
Hypodermic Medication.
  By Bourneville and Bricon.

Practical Points in the Management of Diseases of Children.
  By I. N. Love, M. D.
Neuralgia.
  By E. P. Hurd, M. D.
Rheumatism and Gout.
  By F. Le Roy Satterlee, M. D.
Electricity, Its Application in Medicine.
  By Wellington Adams, M.D. [Vol.I]
Electricity, Its Application In Medicine.
  By Wellington Adams, M.D. [Vol.II]
Auscultation and Percussion.
  By Frederick C. Shattuck, M. D.

SERIES V.

Taking Cold.
By F. W. Bosworth, M. D.

Practical Notes on Urinary Analysis.
By William B. Canfield, M. D.

Practical Intestinal Surgery. Vol. I.
Practical Intestinal Surgery. Vol. II.
By F. B. Robinson, M. D.

Lectures on Tumors.
By John B. Hamilton, M. D., LL. D.

Pulmonary Consumption, a Nervous Disease.
By Thomas J. Mays, M.D.

Lessons in the Diagnosis and Treatment of Eye Diseases.
By Casey A. Wood, M. D.

Diseases of the Bladder and Prostate.
By Hal C. Wyman, M. D.

Artificial Anæsthesia and Anæsthetics.
By DeForest Willard, M. D., and Dr. Lewis H. Adler, Jr.

Cancer.
By Daniel Lewis, M. D.

The Modern Treatment of Hip Disease.
By Charles F. Stillman, M. D.

Insomnia and Hypnotics.
By Germain Seé.
Translated by E. P. Hurd, M. D.

BOOKS BY LEADING AUTHORS.

SEXUAL IMPOTENCE IN MALE AND FEMALE.......... $3.00
   By Wm. A. Hammond, M. D.
PHYSICIANS' PERFECT VISITING LIST 1.50
   By G. Archie Stockwell, M. D.
A NEW TREATMENT OF CHRONIC METRITIS .... .50
   By Dr. Georges Apostoli.
CLINICAL THERAPEUTICS....................... 4.00
   By Dujardin-Beaumetz, M. D.
MICROSCOPICAL DIAGNOSIS...................... 4.00
   By Prof. Chas. H. Stowell, M. S.
PALATABLE PRESCRIBING........................ 1.00
   By B. W. Palmer, A. M., M. D.
UNTOWARD EFFECTS OF DRUGS 2.00
   By L. Lewin, M. D.
SANITARY SUGGESTIONS (Paper)................. .25
   By B. W. Palmer, M. D.
SELECT EXTRA-TROPICAL PLANTS.... .... 3.00
   By Baron Ferd. von Mueller,
TABLES FOR DOCTOR AND DRUGGIST 2.00
   By Eli H. Long, M. D.

GEORGE S. DAVIS, Publisher,

P. O. Box 470   Detroit, Mich.